THE NATION AND THE IDEAL CITY

STUDIES IN THE SOCIAL SCIENCES

edited by C. A. O. van Nieuwenhuijze

1

THE NATION AND THE IDEAL CITY

THREE STUDIES IN SOCIAL IDENTITY

by

C.A.O. VAN NIEUWENHUIJZE
Brooklyn College

1966
MOUTON & CO.
THE HAGUE · PARIS

Printed in The Netherlands

CONTENTS

INTRODUCTION

It seems unnecessary to send a volume on nation, nationality and nationalism on its way with apologies for adding a bucketful to the already swollen river of publications on the subject. Clarification on matters of topical, indeed vital interest demands a cumulative effort. We have lost the 19th century habit of calling a book a "contribution to the study of . . .", but the idea behind the term is still valid.

To study the nation and what pertains to it means to engage in a struggle on two levels. One is the level of conceptualizing and of theorizing, whether in terms of any one of the social sciences or otherwise: the attempt at intellectual mastery over phenomena. The other is the level of interaction with reality: the effort to play one's role consciously so as to exert an impact that will help to mould things in a particular, desired way. The two levels are constantly interlinked; but he who short-circuits them risks to find himself caught in confusion thus caused. Even without such confusion, there is much at either level that remains unclear. Is it because we are too deeply engaged to achieve the mental distance necessary to really see what we talk about? Is it because our theoretical terminology has insufficiently liberated itself from the flexible polyvalence of common speech?

It is not customary to fill forewords with anecdotes concerning the manner in which the book was written: acknowledgments is all the reader is allowed to see. Even so it seems more than purely incidental that part of the first essay was written on a trip that took the writer through Nicosia, in September 1964. Capital of a recently independent country, strife-torn and giving rise to doubts about the future prospects of its independence. Island inhabited by Greeks and Turks and, as somebody has put it, "none of them Cypriots". A lady who boarded the plane told of three weeks of entirely peaceful holidays; she thought the war must be fought mainly in newspapers. And as for the plane, it carried the name of a British as well as that of a Cypriot airline. Could one want more convincing proof that not all is clear today in regard to nations and nationalities? (And, for that matter, could a writer on the subject feel more stimulated to go on with his writing?)

The three essays represent three ways of applying the principles of relativity to the subject matter concerned. The first seeks to clarify it through placing it in the perspective of a typology of the modalities of human togetherness, complemented by sketches of the two main typical, and one a-typical manifestation of the complex under review. The second continues this effort at clarification through relativization by means of pointing to alternatives to "the nation" for some of the major uses that are made of it. Particularly lesser sociocultural entities are considered in the connection. The third essay complements the second in that it considers the perspective of thought, image formation and ideology that is one of the corollaries to the existence of sociocultural entities, including nations. Again a typological approach is employed. This three-pronged exercise, it is hoped, will result in the understanding that the nation is not a unique phenomenon but one out of many forms of human togetherness, and in an additional understanding of what are its distinctive features.

In order to employ the typological approach to the subject it has been necessary to indulge in model building and theorizing to an extent. To some readers this will represent an interest for its own sake. They should be prepared to take the discussion of the nation as a fair case in point allowing the exercise in model building; perhaps whilst conceding that the matter has more sides than this.

I

THE NATION – SOME AFTERTHOUGHTS

1. Gregariousness is one element of the human condition. Biologically as well as intellectually, man finds himself part of a context of which plurality of humans is a major determinant. Human reflection upon these basic facts of life is thus necessarily a posteriori. Like in the book of Genesis, where man gives names to all things after they have been created: naming, and ensuing reflection, are a *prise de conscience* – they are definitely not the creative act. There is thus in principle a difference in kind, an incongruence, between the given fact of human gregariousness and any names given to, or reflection upon, this gregariousness.

1.1. The epistemological side of this difference, usually singled out for further analysis, is of no concern to present purposes. The point here is that any cogitation or arguing on the basis of existing names presupposes, yet does not necessarily account for, the fundamental incongruence between name and phenomenon named. Worse, by failing to account for the difference it risks begging the question on the most fundamental issue, namely the implications of the inevitable incongruence between name and phenomenon named: in other words, the question concerning the degree to which name is adequate to phenomenon named.

1.2. When we speak of "society" we normally suppose that the name covers the reality thus named and in talking about society we normally proceed upon that assumption. In so doing we usually omit to note that it is in its turn normal and by no means exceptional that we should run into difficulties such as are often described as matters of definition. Any such difficulty is at root nothing but a reminder, duly served to us, of the fact that our assumption is subject to qualification. The name of a thing (phenomenon, instance of reality), as compared to the thing itself, is in a way ad hoc. In a manner that qualifies its validity, it is conditioned by contingencies inherent in the act (or process) of naming

– circumstances of perception, established (naturally selective or eclectic) modes of cogitation (also called patterns of thought), incidental needs of various kinds.

1.3. Accordingly, the use of a name in attempting to grasp the reality thus named is a two-pronged operation. On the one hand it is the effort to establish a working relationship between mind and reality (more precisely, instance of reality) concerned. On the other hand it is an exercise in the substitution of an indicator – the name in question – for the reality concerned. The latter operation should necessarily be conducted under high critical awareness, but it seldom is. The less critical consciousness is involved in its execution, the more will it constitute a hazard to the former operation. Any time we discuss society in a manner of unconcern about the meaning of the word "society", we are in for confusion as regards society itself and how it works.

2. These introductory considerations of a more or less general nature are abundantly exemplified in the – exasperatingly vast – literature concerning the present subject: the complex of nation, nationality and nationalism.

Much ink has flown in attempts to grasp phenomena coming, one way or another, under this threefold heading. But the frequent use of the terms seems to have contributed not so much to clarity as to poly-interpretability, if not outright confusion. And why? The answer is, of course, not always the same. A recurrent answer is that people take the names for granted to the extent of applying them inconsiderately. In their use of the terms, they show complete disregard of the fact that these terms, like so many others, are *situés et datés*: and this in an emphatic, indeed dramatic, manner. This careless use sets out from adopting such meaning as the terms concerned had at the moment of becoming fashionable, and proceeds by expanding this from one social and geographical context to others and from one period towards later, as well as towards earlier, periods. It often happens that terms are thus treated as if they were elastic; the treatment concerned has been duly provided with beatific names such as generalization or abstracting use of a word. No doubt, the margin between scholarship and empty verbalism remains thin: a condition that could hardly be beneficial to the humanities at large or to the social sciences in particular.

3. Under the circumstances, the present discussion may steer either of two courses. Either one may set out from a profile of nation, nationality

and nationalism in their heyday in Europe and expand from there: on the one hand into the customary analysis of features and on the other hand into the survey of precursor and successor phenomena and of comparable phenomena in other contexts. Alternatively one may set out from a provisional approach to that of which the complex concerned is one incidental, albeit quite notable, manifestation: thus placing the concept concerned in its proper perspective and thus again better equipping oneself for any comparisons that may be called for. The former procedure is necessarily burdensome as it would involve a critical dialogue with countless authors at every single step. The latter offers more latitude of free movement and seems therefore to be preferable for present purposes.

Thus approached, the present subject matter will appear as the modalities of human togetherness (this being a better word, in the present connection, than gregariousness): with particular reference to one such modality, commonly indicated by the terms nation, nationality and nationalism.

4. Modalities of human togetherness have at least two more or less neatly distinguishable qualitative aspects. (1) They are determined by some particular configuration of data that may be either (historically or analytically) preexistent to them (what the Germans call *vorgefunden*) or coexistent with them if not inherent in them. (2) They are, or tend to be institutionalized, and thus virtually self-perpetuating, in some particular way. Furthermore, (3) they will be distinct according to scope or order of magnitude.

Once determinate in these three regards, a particular mode of human togetherness may be counted upon to be manifest in specific, given cases. A particular mode of human togetherness may, on this account, also be called a particular kind of sociocultural entity. Without getting involved in a philosophical discussion of Weberian ideal types, one may summarize the present argument by saying that a given sociocultural entity, whilst having its own specific identity, must, in order to appear as intelligible,[1] at the same time be a manifestation of a particular kind of sociocultural entity, a particular mode of human togetherness.[2] This view will find further substantiation and clarification in a – necessarily sketchy – elaboration of the three points just listed.

[1] Whether for purposes of consciously living it or of analyzing it in the manner of the social sciences.

[2] It can of course be argued that to speak of specific identity in the connection

5. First now some remarks on preexistent, coexistent and/or inherent determinants.

5.1. One example is found in that complex of phenomena which constitutes the natural setting of life on earth, including human life – the subject matter of ecology.

5.1.1. Referring to "the accumulation and comparison of ethnographical records" C. Daryll Forde (1961, 2f.) feels that "we advance slowly towards a fuller understanding of the underlying factors and processes". He says,

> there has been explanation by the physical environment, that is, according to differences in climate and vegetation which lead men to one kind of activity, type of society, and even of religious belief in one region, and to different activities, social institutions and beliefs in other regions. Despite the intimate relation between human activities and the conditions and resources of the physical world, there are clear limits to this explanation, for in regions closely similar in relief, climate, and vegetation sharply contrasted types of human life are to be found.

This approach sets out, most typically, from the observation that human togetherness occurs in the plural and that its plurality implies differences between its various instances. The two sets of facts are then explained by reference to the kind of factors in question. The approach employed here aims directly at the specificity of any instance of human togetherness: thus implying, tacitly, the plurality and divergence that get the limelight with Forde. Even so, the factors he mentions remain fully relevant to present purposes, as determinants of specificity.

5.1.2. Yet, if it comes to envisaging in which way these factors will exert a determining impact, Forde's elaboration of his theme is perhaps less effective than Leslie Reid's (1962) discussion of habitat and related concepts: with reference not so much to humans as to plants and animals. Reid's approach has the additional advantage that it does not purport to explain one thing by another, which always begs the question as to the explanation of the other. Instead, it handles two orders of phenomena (and of concepts by which these are indicated) in such a manner that their empirical interaction (and conceptual inter-

---

refers to, but does not fully account for, the uniqueness of a given sociocultural phenomenon. If this be correct, one can but conclude that the necessary urge to link universals with particulars meets a limit, on the side of the particulars, beyond which it can perhaps not go. Could it be that this marks the real divide between the method of the arts, in particular of history, and that of the social sciences? (Comp. my *Society as Process,* The Hague, 1962, Ch. IV).

relatedness) become clear. This is precisely what was intended above by introducing the term determinants.

(. . .) the whole earth is a habitat fitted for living things, a habitat of infinite diversity, yet at the same time dovetailed into basic unity. This means that the one great habitat which is the surface of our planet is divided into smaller, but still from our point of view very large, natural regions, each of which is a major habitat with plant-forms and animal-forms characteristic of it and adapted to its conditions. Each is a dwelling-place with its rightful inhabitants, and these are of two kinds, a plant community and an animal community. (p. 38)

Referring to differences in relief or slope, with their consequences for temperature, humidity, rainfall, pressure and soil conditions, the author says,

All these help to demarcate minor regions within major ones, smaller habitats, each with its plant and animal community, within bigger ones. (p. 48)

Quite typical is the following,

It is of the utmost importance to lay stress on the essential unity of any one habitat from the largest to the smallest. Climatic conditions and soil conditions, intimately bound up with one another, are responsible for the habitat being there in the first place, but it would be a capital error to think of it as a sort of container into which the plants and animals are introduced as its inhabitants. The plants are just as much a part of it as its climate and soil. So are the animals. All four of the components are inextricably linked. They *are* the habitat (. . .). (p. 59)

It is again significant that the author should, for purposes of further elaboration, have chosen to employ concepts such as strands of dependence. In this elaboration terms like niches and territories feature along with notions such as food chains and considerations of size, number and density. (As to the latter, some reference will be made to them below.)

5.1.3. No doubt the elaboration of these ideas for the plant and animal world poses problems considerably less difficult than would the same for the human world. Ecologists setting out from this viewpoint have a surer grip on their data than comparable ethnologists have on theirs. Also the latter are subject to more temptations. Just take the case of geopolitics. It represents much the same explanatory approach to differences between (a particular kind of) sociocultural entities as Forde's. More specifically, to differences as apparent in (particular kinds of) interaction between such entities. But its application – not necessarily by Nazis only – becomes suspect any time explanation of past events is stretched into prognostication of future ones and, worse still, into justification of present ones: by reference to some iron law of nature allegedly governing the lives of human communities.

5.1.4. Lest one fear that all is dangerous that occurs in the connection, one should recall that the same trend of thought is represented in the ecological school of sociology and in its progenitor, social geography. On the other hand, most of the works produced in these schools are monographic: they could not be expected to broach the matter of sociocultural entities per se. Rather they will treat these as the, more or less tacitly presupposed, contexts in which feature the phenomena to the description and perhaps analysis of which they are given up. Numerous studies of the Chicago school exemplify this trend (for a brief discussion see John Madge, 1963, 90ff.).

5.2. A second group of preexistent, coexistent, and/or inherent determinants of human togetherness may be summed up as culture: the term here to be taken in the sense of network of meaningful relationships, both between men and between man an nature. What a particular (persistent) instance of human togetherness *is*, is primarily that what it *means* to those who are part of it (and, to a much lesser extent, also to those who are not).

5.2.1. This view has been most convincingly presented and vindicated in the writings of anthropologists. One example out of many is provided by L. Lévi-Strauss' short description of Caduveo and Bororo cultures and societies (1955, Ch. XX, XXII, XXIII). Even so, to speak of culture at large in the connection is somewhat embarrassing. It could, in fact, jeopardize the clarity and simplicity in understanding that it is here sought to establish. Culture is one of those concepts on which there has for a long time been, and still persists, rather too much confusion. With the result that everyone feels free to go about it in the manner that suits him best. (For example, J. S. Slotkin, 1950, Ch. IV). A fair indication of what is here meant may be found in the following statement by David Bidney (1953, 127) here quoted somewhat out of context,

(. . .) by "culture" in general, or human culture, we refer to the abstract, morphological character of the culture of man considered as a species of animal; by "a culture" we refer to the specific modes of behavior and thought, of theory and practice, of social ideals and institutions, together with the products of these cultural activities which are shared in whole or in part by the members of a given society.

The same author (p. 117) points out that the main difficulty resides in what he calls the tendency, amongst anthropologists, to identify the social and the cultural. Indeed the two notions have too often been defined in so broad or crude a manner that they have ended up being practically undistinguishable. With this caution in mind, one can nevertheless trace the point made here in the writings of most of the major authors in the field. Thus, for example, Clyde Kluckhohn (1950, Ch. II) voices the same warning against confusion of culture and society (p. 31) and then proceeds,

> A "society" refers to a group of people who interact more with each other than they do with other individuals (...). A "culture" refers to the distinctive ways of life of such a group of people.

Furthermore, he observes,

> The principal claim which can be made for the culture concept as an aid to useful action is that it helps us enormously towards predicting human behaviour (...). Knowledge of a culture makes it possible to predict a good many of the actions of any person who shares that culture.

Taking prediction *cum grano salis*, as the context in which the word is used seems well to warrant, one can read this statement as arguing that culture determines the particular modalities of a given instance of human togetherness. The same in yet a different way emerges from the following,

> Cultural relativity means (...) that the appropriateness of any positive or negative custom must be evaluated with regard to how this habit fits with other group habits. (p. 46)

5.2.2. A considerable part of the difficulty just mentioned disappears if one considers one feature of culture, rather than culture at large, as a determinant of the modalities of human togetherness.

5.2.2.1. Thus, language as a determinant of major units of (a particular kind of) communication; with dialects, jargon etc., as determinants of sub-units and with language families and the like referring to (actual or virtual, or perhaps past) super-units. Without necessarily having to agree with Bloomfield's (1935, 42) statement that the speech community is the most important kind of social group, one can easily envisage that there must be consequences, in terms of determination of the modalities of communication and thereby of human togetherness as such, to the fact that a speech community is formed by those using the same system of speech signals (*ibid.*, 29).[3]

[3] A useful pointer to the interconnectedness of the several kinds of determinants

Interestingly, it is the anthropologist and the anthropologically oriented linguist who contribute most to studies of this subject matter. But there is usually more stress upon the implications of linguistic differences for observation habits, thought patterns, selectiveness involved in communication, than upon the sociocultural entity proper of which all these establish as many essential features. (See, for example, Kluckhohn, 1950, Ch. 6.)

5.2.2.2. Like language, religion may be considered as one culture trait that will act as a determinant of the modalities of human togetherness. A work like Dunlap's *Religion, Its Functions in Human Life* (1946) comes near to expounding this view: at least in its title and Preface,

> In order to understand the role a religion may or may not play in the civilization of the future, it is necessary to understand the roles that the religions of the past (. . .) have played in the cultures of which they are integral parts. (p. v)

However, most of the book is given up to a discussion of religion per se; the section on "The Social Applications of Religion" (p. 299) in particular is somewhat disappointing for present purposes. Considerably closer to what is intended here comes Joachim Wach's *Sociology of Religion* (1947), which is given up to an investigation of the sociological function of religion. Its integrative impact, later to become the focal point in a study of Islam by Watt (1961) is submitted to consistent analysis. Organization of and differentiation in society, as well as the interaction between religion and the state, constitute the main parts of this study.

> The sociologist of religions will have to study and to classify with care the typologically different organizational structures resulting from divergent concepts of religious communion. (p. 34)

Studies that implement the viewpoint here described vary as to the particular direction they take. Like the work by Watt already quoted, but with a theological rather than a culture historian's interest, L. Gardet's *La cité islamique* (1954) envisages the implications of the basic data of Islam for the Islamic society and policy. (The latter by itself being, no doubt, subject to varieties of manifestation, as discussed in a volume of essays edited by G. E. von Grunebaum, 1955.)

---

here listed can be found in such writings as Sapir's (1949, 89 ff.) paper on "Language and Environment". The article is an excellent lesson in caution in regard to the matter of culture and physical environment.

The analytic approach that is characteristic for works like those by Gardet and Watt, although it may at times be open to questions of objectivity, is clearly less involved with normative preoccupations than is the case in a work like that by Ernst Troeltsch (1931) in which the study of *The Social Teaching of the Christian Churches* is specifically relevant to the conditions prevailing in the Christian community of which the author is a member himself.

More specialistic elaborations occur as well. Thus the study of particular teachings of a particular religion, with normative rather than analytic aims, which F. H. Knight and T. W. Merriam have published under the title *The Economic Order and Religion* (1947). Most famous of all studies of this type (and no doubt much less normative than the one just quoted) is Max Weber's study of the Protestant ethic. A particular type – or should one say, a particular development – of society is, in this work, ascribed to a particular (development in) religious outlook. The seminal importance of the work derives, to a considerable extent, from its having inspired many attempts to approach the development of various kinds of society along basically the same lines. Thus far, the repute in which Weber's theory is held has suffered remarkably little from the refutation it has met at repeated occasions.[4] A less known work that follows, in a way, in Weber's traces is Hoffet's (1948) attempt at understanding American society from its protestant roots.

As compared to that of language, the analysis of religion for present purposes undoubtedly poses problems of considerably greater difficulty. It is less distinct from culture at large; its manifestations are less readily accessible to conceptualization and analysis.

5.2.2.3. Again more difficult is a third feature of culture that may be mentioned in this connection – technology. It is hard to isolate technology from culture at large. To assert its significance as one determinant of the modalities of human togetherness is accordingly hard. Even so one has no difficulty to envisage that a human community that employs the wheel is an altogether different proposition, in regard to its natural context as in regard to its proper functioning, than one that does not. The same goes, a fortiori, for communities with or without devices for amplifying muscular power or for energy production, with or without devices for storing and, more important, accumulating (and also testing) knowledge, with or without devices for mediated communication.

[4] See for example N. Jacobs (1958), K. Samuelsson (1961), and R. W. Green, ed. (1959).

The study of technology otherwise than in order to trace its development in a context of history of culture, has for a good part been channelled into a direction that is only distantly related to the present subject. It is fashionable nowadays (fashion to which Unesco and affiliated agencies have contributed in a measure that remains to be vindicated by the proven importance of the approach concerned) to consider technology as a determinant, not so much of the specificity of human societies as of their development. Fundamentally, there is not such a deep gulf of difference between the two approaches. In practice, however, the difference is more than ample to render most of the studies carried on under this banner irrelevant, or almost so, to the present subject.[5]

Much more important from the present viewpoint, although as yet difficult to assess for the endurance of their significance, are works like those by Freyer (1963) and Ellul (1954). By the manner in which they argue that technology engenders an entirely novel kind of society, they have relegated to the shelves an entire, quite respectable, literature on technology as one, somehow disquieting, feature of civilization: *pace* Ortega y Gasset and so many others.

6. The second category of determinants of the modalities of human togetherness has been summed up, above, in the term institutionalization: this taken to imply self-perpetuation. Institutionalization could be provisionally indicated as the coagulation of numerous data or factors of the kinds indicated just now, into distinct constellations equipped with the means for self-perpetuation.

6.1. In keeping with the traditional preoccupation with abstractions of more static – not to say absolute – connotation, sociologists have tended to devote more attention to institution than to institutionalization. Typical in this regard is James K. Feibleman (1960) who considers the institution as "the isolate peculiar to sociology" (p. 21): to the extent that to him sociology is the study of institutions (p. 19). On the matter of institutionalization, or emergence of institutions, he restricts himself to some fairly simplistic statements on needs, inquiry and beliefs as leading to institutions, and furthermore to a distinction between what he calls the two properties characterizing institutions, to

[5] See my paper "Technological Change as a Factor in Development", to appear in a volume of essays entitled *Intelligible Fields in the Social Sciences* (The Hague, Mouton, 1966); also for further references.

wit social group (social order) and material tools (artifacts) (p. 22).[6]

6.2. More helpful for present purposes is a Dutch sociology text (Doorn en Lammers 1962) in which institutionalization is discussed per se. The authors start out from a distinction between culture and structure which to them is fundamental to sociological reasoning. It is therefore likely that one's approval of their views on institutionalization will be qualified according to one's acceptance or rejection of the underlying basic distinction. The views underlying the argument here developed thus far are definitely at variance with the way in which Van Doorn and Lammers handle culture and structure. Even so, their treatment of institutionalization remains acceptable to a considerable extent. They see a process of emergence of structure in the interaction and communication of individuals in a common setting (p. 119), leading to a measure of persistence (p. 123). As conditions to such an occurrence they list a certain measure of frequency of interaction (p. 124), coordination of interaction (p. 126), the occurrence of an occasion (they speak of an urgent problem occasioning group formation – p. 127 –, which is perhaps saying rather more than one has to), a common cultural frame of reference (p. 129) and leadership (p. 130). It is particularly in the listing of conditions that one may wish to qualify the concept: as stated, according to one's acceptance or rejection of the authors' premisses. For example, to the present viewpoint it should be unnecessary to list the common cultural frame of reference, this having been considered already as one of another kind of determinants of the modalities of human togetherness. In the same manner, leadership could be censured as too crude a term, in need of being replaced by some notion less anthropomorphic and more adequately indicating the focussing process concerned.

6.3. By way of further indication of what is here meant two further references are apposite, each to an approach that covers part of the ground. Relevant to what may be called incipient institutionalization is a phenomenon briefly discussed by Feibleman (1960, 116), the aliena-

[6] For purposes of fair rendering, the following quotation must be added. "If, as we have said, an institution is that subdivision of society which consists in human beings in groups established together with their customs, laws and material tools, and organized around a central aim or purpose, then there are six necessary elements which can be counted as parts of every institution. These are: *social group, establishment, customs, material tools, organization* and *central aim*" (p. 23). Compare also B. Malinowski (1944). For present purposes it seems convenient to avoid a discussion of Parsons' (1951) presentation and elaboration of the concept.

tion of the artifact. He merely drops the clue, but it seems worth pursuing. Relevant to what could be styled, in its turn, achieved institutionalization are, in the first place, most of the studies on institutions. (One good example out of many is Berelson and Steiner 1964.) In this regard one should keep sight of the useful distinction introduced by André Marchal (1959, 144, 150), between "institutions-choses" and "institutions-groupes". More in particular, one may mention studies concerning the image,[7] such as launched by K. Boulding (1956). In a way, these always refer back to *Gestalt* psychology (comp. Wertheimer, in Ellis 1955).

7. The third category of determinants of the modalities of human togetherness seems to stand a bit apart from the other two. Whereas these appear as operational in such a manner as to have qualitative consequences, this one appears as quantitative yet relevant for its qualitative implications. Particularly economists have considered the implications of size, scale, scope, or order of magnitude for the nature of economic phenomena. Thus, the 1957 conference of the International Economic Association has been given up to the *Economic Consequences of the Size of Nations* (Robinson, 1960). Besides economies of scale, as a problem in size and efficiency (Jewkes, *ibid.*, 95ff.), they have discussed size in regard to stability (Jarshish, *ibid.*, 190ff., also Leduc and Wieler, *ibid.*, 200ff.) and in regard to viability (Rothschild, *ibid.*, 168ff.). Closer to the present subject of the overall characteristics of a given sociocultural entity, although considering them from a particular angle, is Kuznets' discussion of the economic growth of small nations (*ibid.*, 14ff.). The same author has contributed a paper to a conference volume entitled *Challenge of Development* (1959).

7.1. There can be little doubt that political scientists and sociologists, to mention only these, stand to benefit by an approach in terms of size to the matter of specificity of sociocultural entities. Thus far, this has been somewhat neglected.

In a way the matter of size is accounted for in the differentiation between micro and macro levels, both in economics and in sociology. In sociology, this had resulted in problems, notably the issue of the middle range.[8] In economics on the other hand it has resulted in a

[7] On the metasocial aspect of the image see below "The Ideal City or the Varieties of Metasocial Experience: A Typology".

[8] Comp. my paper "The Quest for the Manageable Social Unit: Is There a Middle Range?", to appear in volume referred to above. References are given there.

separation of what are virtually two sub-disciplines. The main difficulty is that this distinction is too crude to do justice to size: this being a matter of plural not of binary manifestation (comp. Leent, 1962).

8. Given these three ranges of determinants of the modalities of human togetherness, one is led to conceive of any configuration of their incidental impact (or effect) as one resultant complex. This in its turn may be fittingly described as an emergent sociocultural entity.[9] Conceptualization in this regard is helped along by studies already made; their immediate concern may to an extent be at variance with the present one without impairing their relevance to the argument here developed.

Abraham Edel, in a paper on "The Concept of Levels in Social Theory" (Gross, 1959, 167ff.), concentrates upon the perspective of the time dimension involved in emergent levels: or, as he calls it, the developmental (p. 168) or historical (p. 130) perspective. His further concern is with the range of levels (superimposed rather than juxtaposed), for the analysis of which he introduces the concept of locus (p. 172), thus neatly relativizing the matter of plurality and that of complexity both in one stroke. A further virtue of his treatment is the suggestion that levels may be conceptually determined in more than one way and in fact simultaneously in various ways. This would seem to tally with the view developed above, of various sets of simultaneously operative determinants. Leaving aside, for a while, Edel's concern with superimposed levels and keeping his dynamic, operational approach, one envisages the possibility of conceiving of sociocultural entities as emergent (the word taken in a sense implying a span of self-perpetuation or of self-sustained operation) from the joint effect of what, if considered in the "opposite" direction, features as ever so many determinants occurring in incidental configuration.

8.1. What has thus far been said becomes perhaps clearer in pictorial representation. The three orders of determinants discussed can be shown each as one plane. (This is not the occasion to discuss how one could use the possibilities of the plane so as to achieve a more or less adequate presentation of the inherent variabilities of each.) A particular configuration of instances of determinants – that is, a particular emergent sociocultural entity – will thus be represented as the

[9] The same, considered from the angle of its permanence, could be called social (or sociocultural) identity. Comp. a paper under this title in my *Cross-cultural Studies* (The Hague, Mouton, 1963).

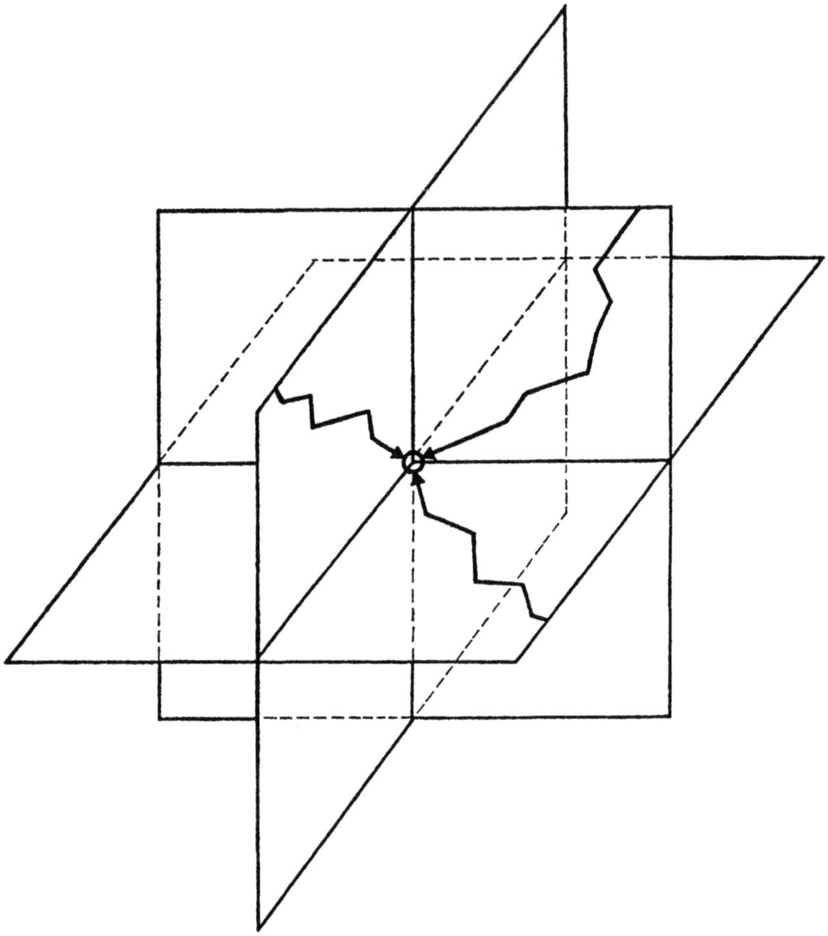

Fig. 1.

intersection of the planes. So long as one employs three ranges of determinants a neat graphic presentation is possible. With four or more, the principles of presentation remain the same but one could not so easily draw it (see Figure 1).

Imagine three pieces of glass that intersect. Each has a crack, representing a graph of the particular determinant of that plane; and the three cracks intersect at the point where the three planes intersect.

8.1.1. Note that the intersection of planes and of graphs on planes at one and the same point is not a coincidence. In principle each plane is, so to speak, mobile along its own third dimension. (Which however

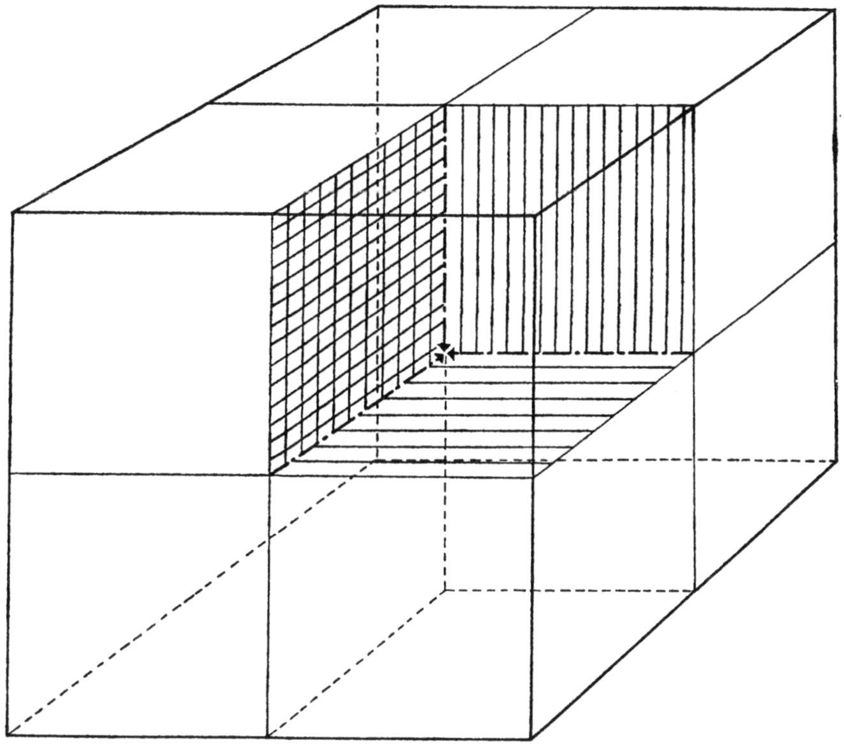

Fig. 2.

is not shown but by implication, inasmuch as it coincides with the intersection line of the two other planes. The picture, whilst actually showing three dimensions, implies a fourth.) One draws the planes where the lines occurring in each of them intersect. The same applies in a two-plane drawing like Arthur Koestler employs to show his idea of bisociation (1964, 35 et passim). If this is felt as a difficulty it can be easily avoided, though at the cost of neatness in showing a given intersection point, a given emergent sociocultural entity. All one has to do then is turn the open three-plane drawing into a closed one (surely without difference in meaning), thus making it a cube. At any point of the hollow cube one could pinpoint a "locus": the intersection point of the three orders of determinants each in a given instance. If the cube were of wire one could hang there a marble or so, with the name inscribed on it. On each of the three planes that face the eye (the other three are mere reduplications), two intersecting ribs (the other two being again reduplications) would feature as co-ordinates for graphic presentation of the determinant concerned (see Figure 2).

9. It has now been shown how modalities of human togetherness can be envisaged as being determined in several respects. It has also been suggested that a particular modality of human togetherness can be understood as a configuration of specific instances of each of the several determinants mentioned. The configuration in its turn is not so much one specific sociocultural entity as one kind of sociocultural entities: the former having to be seen as one instance of the latter – a real instance of a conceptual category.

9.1. The emergent sociocultural entity does to the writer's knowledge not occur under this name in the literature. On an earlier occasion (1963) I have indicated much the same thing as social identity. But the two are not quite the same. The latter term indicates a specificity that is lacking in what is denoted by the former. In order to trace this specificity one must not merely analyze the modalities of conditioning, as described above, to several ranges of determinants. In addition one must look for operational modalities, as pertaining to any emergent sociocultural entities: that is, to any configurations of instances of the several (conditioning) determinants discussed.

One is perhaps tempted to say that this means adding a dimension to the relatively neat picture first drawn. As a first approximation this is certainly a useful manner of envisaging the situation. Upon closer inspection it is clear that the range of (operational) modalities that enters now upon the scene, must in its turn refer to determinants of its own.

9.2. The question is then which are these determinants. It would seem that at least three of them can be distinguished, along with a fourth that need not be given a place in its own right inasmuch as it is somehow correlated to one of the other three.

A brief listing is all that can be provided here, and this on a purely tentative basis. The purpose of this paper is not model-building, let alone elaboration and corroboration of models. It is just that for the purposes of the present subject models of this nature are a necessity in order to assess the phenomenal context in which nation, nationality and nationalism occur. An exercise, in other words, in sophistication: meant to get away from distorted and value-ridden presentation of the concepts (and phenomena) concerned.

The following, then, can be listed as determinants of the operational modalities of social identity, that is, of the specificity of emergent sociocultural entities.

(a) The comprehensiveness of the sociocultural entity concerned, as related to its institutional frame of reference. In principle this is a ratio between two variables, but the variables are not easy to pinpoint.

(b) The range of significances that integration of the emergent sociocultural entity has for its self-perpetuation. This range can be described as a mix of inner-directedness and outer-directedness in the mode of self-identification, with a shifting degree of preponderance that either of the two has over the other. Inner-directedness is used in a sense that connotes self-realization of the entity concerned, for its own sake: per se. Outer-directedness on the other hand connotes distinctness of the entity concerned as regards other entities occurring in the universe in which it exists.

(ad b) Another range could perhaps be substituted for the one mentioned under (b), as a determinant of operational modalities. It sounds rather different but it would seem that the two are correlated one way or another, so that it is sufficient to employ only one of them in the model. This is a range of significances of antagonism or conflict as one operational aspect of the existence of sociocultural entities. One may gain describe it as a mix of two manifestations of different order, with a measure of preponderance of one over the other. One would be conflict as operationally positive, as positively institutionalized; the other would be conflict as operationally negative, as negatively institutionalized.[10] Where the stress is on the latter, and where a sociocultural entity will thus relegate conflict to the informal, hidden or latent side of its operational pattern, it is likely that its self-assertion as regards the outside world will entail a vision in terms of potential conflict, even if this may be appreciated as undesirable. Accordingly, where conflict is positively operative, there will be no such countervailing stress in regard to the distinctness between given entity and outside world.

This is not the occasion to elaborate the correlations between the two determinants here treated as variants of one another: what matters is that there seems to be no reason to operate both at the same time.

(c) The range of rationalizations, current in a given sociocultural entity in regard to its integration. As critical in the connection one may use purposiveness, viz. in setting up a range according to whether it is

[10] This construct presupposes an underlying complementarity construct, to which antagonistic, oppositional, "dysergic" interaction is one element and mutually reinforcing, "synergic" interaction the other. For a presentation of this construct see my *Social Scientists in Pursuit of Social Change* (The Hague, Mouton, 1966).

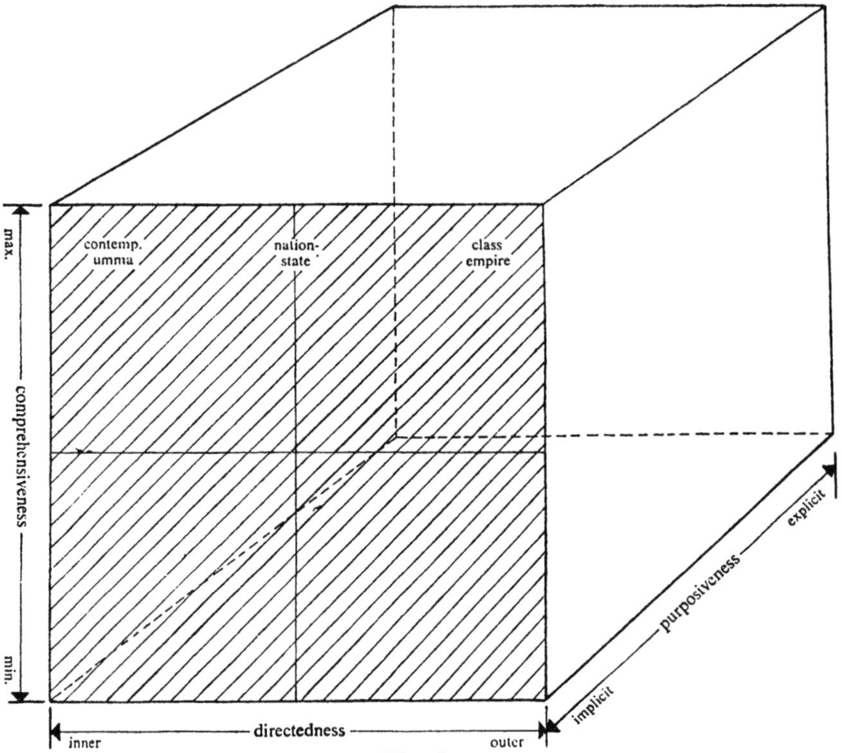

Fig. 3.

implicit, or so to say tacitly presupposed, or on the other hand explicit, a manifest constitutive element.

9.2.1. These three determinants in their turn can be drawn in the form of three intersecting planes or, alternatively, of a cube. To the former three- (really four-) dimensional model, another "dimension" is added, which in its turn is (at least) three-dimensional. Each intersection in the three-plane model turns out to be, upon closer inspection, not one point, but an intersection once over again. Each point in the original cube turns out to be a cube in its own right.[11]

What is perhaps more important is that the "inner" cube has the same virtual content, namely the content it has per se given the nature of the determinants it represents, wherever it occurs in the "outer" cube. But as regards actual content it may vary according to the place where it occurs in the "outer" cube.

[11] The idea of the self-repeating cubes comes from Professor C. H. Van Schooneveld, Stanford University. The particular variant of elaboration employed here is mine.

To give an example. Figure 3 gives an "inner" cube, representing modalities of specificity of sociocultural entities by operational determinants.

The three determinants have been so sketched in that on the front plane of the cube one meets any cases of preponderantly implicit purposiveness. Towards the top of that plane, comprehensiveness reaches the maximum (as related to institutional framework). According to different preponderances of inner versus outer-directedness one can inscribe, in the top of that plane, three names that may exemplify the positions thus signalled (rather than adequately named). To the left, the contemporary *umma*, the world-wide Islamic community devoid of effective organization as such and with but little left of its concern about the heathen world that was typical of earlier days. Towards the right, one may put classical empires such as those of Alexander the Great, of Genghis Khan or of Rome. In the middle one may place the nation state.

Now the point is that if one fixes one's attention for a while upon the position thus (provisionally) labelled as nation-state, as one instance of the construct represented by the cube, and then proceeds to move this cube, so to speak, through the space of the "outer" cube, the relevance of this position may prove to be more or less, according to the place in the "outer" cube where one happens to be. The practical manifestation of this phenomenon is the problem of the viability of states: there are grave doubts nowadays in certain circles whether some of the newly independent states can really exist as such. This then depends upon the other order of (conditioning) determinants; in particular matters of geographical size play a role.

10. In the perspective thus drawn, the complex roughly indicated by the terms nation, nationality and nationalism features as one type of emergent sociocultural entity. Its profile, as a type, can be envisaged as a complex resulting from specification, approximative concretization, in each of the three ranges of determinants listed. What features in general as the question of order of magnitude would, in the connection, manifest itself as a particular, restricted range of orders of magnitude; and similarly for the other kinds of determinants. Complementary specification, applied in its turn to the type thus established, should establish the operational manifestations and thereby provide an even closer approximation to observed reality.

10.1. The nation, or nation-state, has been inscribed in its proper place in the "inner" cube, and it has also been shown that after this it remains to be seen how relevant it is at any given point within the space co-ordinated by the "outer" cube. The nation is an emergent socio-cultural entity of maximal (or should one say optimal) comprehensiveness in respect of its institutionalization, of maximally implicit purposiveness so far as the current rationalizations as to its integration go, and indeterminate in regard to whether the significance of its integration is inner-directed or outer-directed.

10.2. Accordingly, the two related concepts may be provisionally defined, so as to further the clarity of the argument that is to follow.

Nationality can be described, briefly, as the institutionalized image of the nation(-state), whether self-image of its members or image prevailing amongst aliens.

Nationalism, in its turn, is the cumulative or increasingly institutionalized impact that this self-image has upon the people who hold it: impact that moreover tends towards maximization.

10.3. The discussion of nation, nationality and nationalism that is due here may thus continue, if so desired, along the lines of the model here suggested, and by way of further elaboration thereof. This approach is certainly worthwhile but it has one weakness for present purposes – it is rather too speculative and it is so broad that one risks losing track of what one is really after. It is therefore proposed to leave the model at the point up to which it has been brought.

Further understanding can be better served by employing a complementary approach. This approach can to an extent be historical. But it will not be description for description's sake. The answers will be sought to two quite specific questions. First, which is the cultural tradition in which the nation stands upon emerging, as a real phenomenon and as a concept? Second, which elements of this heritage have become eliminated, recast or left intact by its emergence? The two answers together should constitute yet another approach to the specificity, this time historical not typological, of the nation and related concepts.

11. The second approach will set out from the experience of human togetherness that is commonly identified with the small communities, long ago called primitive, that are the favorite objects of investigation of anthropologists and of historians of religion.

11.1. In its original, fundamental sense the "we" notion that signals a given instance of human togetherness, stands for no less than a full universe. It embraces space and time alike – all experienced space and all real time. Real time is the present as well as what the Germans call *Urzeit*, primeval: essence or origin identified with "the" past: a category not controlably, i.e. measurably, related to the present. The community's actual way of life is the optimally adequate manifestation of the laws of the universe; actual practice and ideal type or normative image are virtually identical.

Scholars who study this sort of communities have an inclination to repeat, in their own fashion, what the members concerned feel, namely that a full perception of the comprehensive character of the community is a matter of initiation, thus of an elite of some sort. Yet, the comprehensiveness as such is nothing but the corollary of close living in a small self-enclosed universe. The world begins and ends at the horizon, where the sun rises and sets. As the Kabylian farmer said when at a mature age climbing for the first time the mountains at the foot of which his village lay "Oh God, how great is Your world" (Bourdieu, in Pitt Rivers, 1963, 66).

11.2. It is only in the retrospect that a less primitive way of life of larger scope will afford, that one can distinguish connotations to this prime comprehensiveness. And the words that one has to use in order to name them are all shockingly out of place. If the connotations are there, it is often a matter of more or less embryonic manifestation. With this reserve, the "primitive" closed community is – to just make a random list of some of the most salient connotations, – (1) a language unit (non-members are stammerers, barbarians), (2) an economy of autarchic nature (limited exceptions to the rule evoke such names as Argonauts, whether in classical Greece or in the Pacific as described by Malinowski); (3) a church or religious community (whose cult may well demand sacrifice of humans who in one way or another stand apart from normal human existence) with the maintenance of the entire universe as its main function (through ritual actualizing the mythical fundamentals of existence); (4) a state with institutionalized distribution, maintenance and application of power; (5) a full culture pattern; (6) a self-contained structured society (or if one prefers, community). In it, the individual is embedded to such an extent that a full life can only be had as a corollary to full membership. Accordingly, the alien presents a hazard in all respects listed; he is treated as such, with hospitality that is at the same time quarantaine: not in the medical sense, as it happens, but in all other senses.

11.3. Now this – far from poetic – comprehensiveness may suffer disruption. Normally the main symptom is that the balance between single member and embracing community gets lost. Individualism is a sure sign of disruption of closed communities. The next thing to happen is that some of the aspects here listed as virtually distinct connotations of the original comprehensiveness become actually distinct: again in becoming disproportionate. Then follows maximization of one or a few of these distinct aspects or connotations; and one is in for the development of in principle unlimited numbers of distinct variants of human togetherness, distinct types of community or society. Although the further interpretation of this process is beyond the present scope it is necessary to point out that what is corruption and disintegration in the light of the past is necessary further articulation in the light of the present, and, as such, bound to feature as beneficial.

11.4. The most frequent, and given the analysis here attempted also most likely, development is maximization of one aspect along with subsidiary refocussing of a few other aspects upon this one. Moreover, not all aspects listed lend themselves to such maximization. And even those that do will do so by acquiring effectiveness as denominators of social control. Thus, the aspect of power in the sense of brute striking force, that of wealth as economic might, that of religion as a cultic complex distinct from the way of life or culture pattern at large, and that of civilization as a normative way of life.

11.5. Hans Kohn has in one of his works (1946) written the prehistory of nation and nationalism. This is an important and highly stimulating work. It is, however, presupposed to the present argument rather than an intrinsic part of it. One reason is that, if nation is an emergent sociocultural category, Kohn's work does not so much write its prehistory as trace back into history the conditions under which it emerged and the preexistent material – sociocultural and in terms of ideas – out of which it became built. Another reason is that in this particular case the emergence concerned can be envisaged in more than one way. One is that which Kohn has applied: trace bits and pieces of nation etc. in earlier phases of the same culture tradition. The other is to argue that, in a particular way, the nation represents the effort to reconstitute the long lost comprehensiveness of social existence, on a scale and with a degree of consciousness befitting the conditions under which the attempt is made.[12] The two approaches can be marshalled in

[12] The prime community insofar as it features as a frame of orientation for more articulate and at the same time less balanced, less comprehensive entities,

such a way as to complement one another. But whereas Kohn has, perhaps without considering the second, favoured the first, the second is preferred here because it is better fit to serve present purposes.

The point can be made clear as follows. As it has appeared in Europe at one time and as it continues to appear elsewhere as well, with a peak of intensity since World War II, the nation represents an effort to reconstitute the comprehensiveness of – not so much primitive as prime, primordial society (or should one prefer to say, community). Nationalism, in this sense, has a strong element of holism. Nationality is the never quite achieved product of common self-identification.

12. There is no need to relate the facts in detail. Kohn and others have done it. What must be marked is that in Europe, the birth of nation, nationality and nationalism was preceded, and no doubt prepared, by conditions, as described above, where one aspect or constituent of the prime comprehensiveness of society, supported by one or two others, equally severed from the original comprehensiveness, achieves preponderance. In fact, things were by one step more complicated than this. Late Medieval Europe and to an even greater extent Renaissance Europe have as their main feature the simultaneous occurrence of two trends toward maximization of one aspect – of course in competition to one another. On the one hand European society of those days was primarily church, socioculturally effective institutionalized religion. But on the other hand it was a society ruled and held together by power as brute striking force. Subsidiary aspects in either case were power-as-wealth and culture-as-norm. Competition, then, was necessarily most traceable in social control in these two regards.

12.1. It is perhaps permissible to argue that a moment arrives when such maximization, whether simple or twofold on competitive basis, requires for its furtherance the taking of measures that will in the long run undermine it. Amongst such measures is likely to be the increased reliance, of those effectuating the structure that maximizes one aspect, upon the masses. This is in itself a highly interesting development. The comprehensive integrity of the prime community, as was argued above, breaks down at the point where the single community member emerges as a self-purposive individual. From that moment onward, in keeping with increasing size of society coupled to poor communication tech-

---

tends to assume the aspect of a frame of orientation, if not of belonging – in short, of an ideal city. This perspective is so vast that it demands separate treatment.

niques, the role of leadership proves decisively different. Originally one of, and one with the community, the leader or leadership does not feature as the visible embodiment of the community but at ritual occasions, including emergencies. Henceforth, the leader or leadership will feature as the embodiment, the personification of the state (society) concerned with clear maintenance of his or its distinctness as a permanent factor. The virtual identity of community and leadership has made room for a complex construct in which leadership is elite and as such is the society ("l'état c'est moi": but this, most significantly, was not said until the day was almost over), and where the populace are a mass, irrelevant but insofar as they exist thanks to and on behalf of the elite. It is precisely this leadership construct that in the long run defeats itself. A reconstitution of society is due, and it is seldom realized that this calls for a basic re-evaluation of the role of the individual as a member of society. There are indicators, however, for those willing to see them. What else does the preoccupation with liberty signal but exactly this basic issue?

12.2. Remarkably, revision goes on at another vital point as well. It does not necessarily lead in a direction so as to make it converge with what is being done about leadership, individualism and liberty: those imponderables that must needs appear as fundamental any time society is felt to be ailing. This is the matter of prime society being not *a* universe but *the* universe: self-perpetuating, self-enclosing and complete in every respect – cultural, political, economic, religious, social, and what not. This fundamental trait of universalism becomes dissipated in the process described. On the one hand the church, as the true universal community, proves ineffective as such and the competing universe of the Islamic *umma* is too manifest and too close. On the other hand the various sovereign kings find little confirmation for the self-contained perfection of their several realms in the actual state of affairs, where commoners move and trade freely with one another regardless of whether their respective states are at war or at peace.

12.3. When the state of affairs that carries, in these and other problems, the germs of its successor finds its end in revolution, the new state of affairs represents, in more than one way, the attempt to reconstitute the prime comprehensive society. Bucolic dreams like those of Rousseau are significant in this respect, yet they are less important than one or two phenomena that deserve to be singled out for discussion here.

12.3.1. The first is discussed by many authors, under the name of democracy. Now there is no denying that democracy, although not one of the three catchwords of the French revolution, is a good term to render the gist of the revolution: as a political complex. But one should not allow oneself to believe that the term is adequate to the event. The revolution was political and many things besides. In fact, it was primarily the attempt at undoing the eclectic maximization of distinct aspects of sociocultural existence, such as this had become traditional. Undoing that could mean a return to the comprehensiveness of prime society, to an extent, and/or the replacement of one maximized distinct aspect by another. One needs a term that is better fit than the term democracy to indicate the potentialities of the undoing and in addition also the main positive direction the process turns out to take. It is proposed to employ for the purpose, the term integrative society.[13] Politically this means democracy, as an ideal (not to say ideology) and to an extent as a norm for practice. Economically, it means, surprisingly enough given all the talk on universal values and freedom, mercantilism in its various guises. Culturally, it means separatism of linguistic units. Socially, it means plurality rather than dichotomy of classes, increased proximity of classes and an upward movement built into the new pattern: class differences are ad hoc, no longer fundamental. Again culturally, it means that what used to be the privileged domain of those who were the elite (on other counts, as a rule) is not merely accessible to all but a common responsibility to all. (Thus raising, inter alia, the problem of "democratization" in education that thus far nobody has solved.) Religiously, it means secularism. Or does it?

12.3.2. This introduces the second point. Prime society has the uniqueness that pertains to being the universe. In Europe before the French revolution, this uniqueness had changed names rather than fundamental nature. It was called sovereignty, and sovereignty was vested in the king, by divine right. The king as the embodiment and personification of the community has (to the exclusion of all other members of the community who feature as his subjects) sovereignty and is thus heir to the universality that the prime community had, as of rights and as of necessity, inasmuch as through myth and ritual it actually established the universe: time and again. The divine right is thus quite significant: a Christian presentation of a pagan world view, yet ever the same fundamental consideration.

In the new order established after the French revolution sovereignty

[13] Also used by Geertz (1963). Compare also Jacobs c.s. (1964).

undergoes a shift that is at once a step onwards and a step backwards. Backward: once again it is vested in the entire community – the nation as they are now called. Onward: this is so for no reason but that the nation exists and realizes itself, amongst other things in sovereignty. The religious perspective is gone. Secularism, to take the thread again, is in fact merely progressive secularization: progressive yet ineffective. With God out of the picture the way is open for elements of pagan religiosity without the name. Many are the authors who point out how very religious are some of the connotations of the nation-state. Occasionally there is the brute who says the thing openly. A most menacing example of the trend and also of the dangers it entails is of course Adolf Hitler. A very minor and almost amusing example of what is yet ever the same trend can be found in the proposal by Kaplan (1947) to provide American statehood with more elaborate and more overtly religious ritual.

12.3.3. Apart from the religious aspect which remains dubious, sovereignty as an essential feature of the nation-state raises the problem that in regard to the actually prevailing conditions, particularly in regard to communication between nation-states, it is a relict of earlier days and as such in outright contrast with the main determinants of the context in which it occurs. (Even mercantilism is a valid proposition only as long as it presupposes an ever increasing intensity of economic interaction between states: it works as long as it is the exception to the rule.) The difficulties in European unification that have been rampant ever since World War II bear sad witness to the point.

12.4. One more aspect of still the same matter must be marked. The way European nation-states have developed, the holistic longing for the comprehensive integrity of prime society has been engaged in constant battle against the tendency to single out, once more, one distinct aspect of sociocultural complexity, namely power as technological and economic, in short as organizational control. This strife is creative as such, but this is no matter of concern now. What is of interest is one or two of its manifestations: not internal but as regards the outside world. One is the so-called flight from culture said to be typical of modern Western civilization. It is the individual's decision to seek the wholesomeness of life by keeping oneself eliminated from one's own civilization: a very particular kind of tension that in order to become constructive needs the Robinson Crusoe less than the Pilgrim Fathers or the founders of the Rochdale Community. The other is that the Western self-assertion, state-wise no doubt, operates through seeking the challenge of the non-West. Colonialism and impirialism have far deeper roots than mere lust for power or hunger for profit. They relate to the very essentials of the Western way of life. Small wonder that colonialism once finished is succeeded by what mistakenly though understandably is styled neo-colonialism – a formula in the process of being established, that must at once retain to the West the vital challenge that the non-West offers and take care of a new pattern of relationships between what – pure contradiction! – are quite explicitly universes each in its own right, viz. the so many nation-states of the contemporary world.

13. The focus, in pursuing European developments, has shifted. It is time to consider the non-Western nation-states that have come to fill the benches of the UN assembly.

13.1. Colonization occurs to sociocultural entities that, as Bennabi (1954) and others have remarked but many would rather forget, are colonizable. Mostly this colonizability ensues from a state of affairs in which, like when European civilization began, the prime society crumbles down under the weight of tradition and accretions. Once more, individualism and the increasing distinctness of those in control, as of a different order than the commoner, are symptomatic phenomena. A more or less problematic span of control (unfavourable ratio between size and communication), is likely also to play a part.

13.2. The colony, once established, reduces the entire preexisting society, regardless of its inherent articulation, to secondary status: the colonizers henceforth reserving the status and function of elite for themselves. In so doing they achieve a social structure that in their home countries is decidedly of the past, and rejected at that. In the meantime colonies are non-states: up to a limit they are replicas of the political and economic set-up of the home country concerned. Beyond this limit they are mere appendices to the home country. *Patria,* as the Dutch used to say, *la métropole,* in French parlance: curious terms each, and most significant.

13.3. Colonialism, in becoming more efficient and more enlightened, undermines its own foothold[14] and the impact of the greatest war of all is ample to unseat it. In the process many things become loose. For one, a pattern of world-wide relations, in terms of power and of commerce (to mention only these) collapses, leaving the problem of a

[14] See annex at the end of this paper.

successor pattern as a matter equally vital to the interplay of states as it is to each state by itself. Besides, in each colony or agglomeration of colonies the bottom falls away from beneath the political, economic and social structure: and this under conditions that were already marked by profound perturbances. This is worse than revolution, inasmuch as those who henceforth feature as the agents of this turn of affairs, the national leaders as they are usually called, are not the sole agents and in fact are not the main ones. The other agents, the colonizers and what should perhaps be called the world constellation, are not on hand.

13.4. What this drastic change, not merely of conditions but of an entire frame of reference, will mean is in no manner clear at the outset. Back to the comprehensiveness of the prime society, the loss of which was in fact the breach through which colonialism could enter? Fundamental revision of the colonial pattern, leading to another configuration of maximized and subsidiary distinct aspects of the totality of what a sociocultural entity really is? Or, third possibility, continuation in much the same style, with a new elite in the place of the old? In the large majority of cases there are at least two candidates for the role: the remnants of the old elite that was swept from its feet, but never quite obliterated, by the colonial elite; and a potential elite, on the brink between two worlds due to Western education of some sort or another. With three ways open, the chances are only too great that things may go one way whilst it is claimed that they go another. In the emergency, it is more than likely that the first move of those concerned will be to fill the gap: the third possibility listed – but rationalizing it in terms of the first or of the second possibility listed, or of a cross-breed between the two.

13.5. But these things remain to be seen. The only thing visible at the moment of independence is statehood, as embodied in a new government consisting of autochtonous persons. Note that this independent statehood signals that the former colony will henceforth continue at the international plane as an independent state. At the national plane it has no immediate significance that would not be speculative to an extent. This statehood represents the partial achievement of the effort of those who have tried to unseat the foreign colonial rulers during a usually harrassed period prior to independence day. The positive rationalization given to their urge had without exception been the unhampered self-realization of the colonially ruled sociocultural entity. Its very name, nationalism, had been borrowed from the culture of the colonizers. As such it was the most efficient ideological weapon against them.

The empty shell of statehood thus begins by being filled with content of a fairly precarious nature: nationalism. Nationalism not so much as a reflection upon the given existence of a distinct nation but rather as a commitment to nation-building. Statehood for ex-colonies is primarily a provisional mould in which the process of self-realization of the sociocultural entities concerned is offered an opportunity to take place. Independence is a precondition rather than a fulfilment.

13.6. All the rest, let it be repeated, remains to be worked out. There are, however, certain factors that will play a role, whether noticed or unnoticed. Sovereignty is one. In so far as it evokes the integrity of the prime community, which possibly is fresher in people's minds here than it is in Europe, the newly acquired sovereignty will be treasured beyond measure. Thus it could push into the direction of the first of the three trends just listed. "Beyond measure" is a tragically apposite qualification. In most respects, the large majority of newly sovereign states represent drafts made upon the future. Nationhood is amongst these things.

13.7. What is remarkable, however, is that in one regard at least nationhood is a very apposite term. The second main urge after sovereignty, for which national independence movements have invariably stood all over the world, is that towards what was called above the integrative society. But there is as much atavistic longing in this as there is purposeful appreciation for some achievements of the modern European welfare state and of the Russian and American technocracies.[15] Which of the two will gain the upper hand is a question bound to remain undecided for years to come; neither is it likely to be decided, when the time comes, for each of these countries singly and separately.

13.7.1. The point is that here perhaps more than in the Western world the exclusivity that comes with sovereignty is defeated a priori by the very events that conditioned the birth of these nations, and once over again by the conditions under which they venture upon their independent existence. Nothing is so fit to sum up all the complexities and frustrations that are inherent in this state of affairs, as technical assistance: phenomenon doubly enigmatic on account of the impact of the cold war – this in its turn being a phenomenon of an altogether

[15] The term is chosen for purposes of adequate description not so much of what the USA and the USSR are, but of what they would seem to represent in the eyes of the newly independent peoples.

different nature (or is it?) yet dangerously effective in the connection.

14. For further enlightenment it remains now to consider briefly one case that demands attention precisely in that it is a-typical.

14.1. With due allowance for their specificity as a culture area, the Arab countries are reasonably typical for what has just been observed in regard to the neo-nationalism and the newly independent nation-states of ex-colonial and comparable territories. The specificity is largely a matter of cultural history and to a lesser extent of geographical and climatological conditioning. At several occasions, all or part of the area has belonged to large empires. Besides, its Muslim inhabitants, forming the majority almost everywhere, maintain a significant degree of allegiance to the *umma*, the nomocratic universe of Islam, which the empires just mentioned attempted to embody each in its own way. Religiously institutionalized power-as-striking-force, backed up by wealth, has thus been the aspect of sociocultural existence that has tended to be maximized ever since Islam came to put order to the decaying prime societies of desert Arabia. The culture aspect gained the highlight occasionally, for example in the tension between *ʿArab* and *aʿjam* in early Islamic history and between Arab and Turk in the period preceding World War I. It did not achieve persistent maximization. The contemporary states have been installed, in the wake of World War I and to a lesser extent after World War II, in a manner that could not have been more accidental. On the other hand, these frontiers prove hard to dislodge or eliminate. There is much talk about the Arab nation; for all the emotional appeal it exerts, particularly upon those not in power, it is simultaneously the sign of a desire to overrule jealously guarded configurations of entrenched self-containedness, pointedly called independence. At the same time all this serves as a smoke screen beneath which a trend is afoot toward a more integrative society, in the sense used above. Its integrative pattern is unlikely to be effectively contained in, or by, any of the existing frontiers. On the other hand there is no saying what its geographical shape could be once it would attain fuller manifestation.

14.2. For reasons, and in consequence of forces, alien to the area, the state of Israel has been injected into it. Thus far it has remained an enclave that is barely beginning to try and strike some roots in the area itself. The injection proper, a process that has taken about half a century to show its full extent, is an act of despair coupled to utopianism; as such it is beyond good and evil. The agents are the Jewish diaspora; those who did not bodily migrate to Israel often have an allegiance towards it of which one may doubt whether it could be so categorical were they actually to live in the "twice promised land". (The handful of non-diaspora Jews, those who never left or came back much earlier, are not always considered an asset to Israel.)

The Jewry, diaspora or not, have been held together by faith: faith in their being the chosen people. The manifestation of this state of grace, this having been chosen, since it has failed to appear as glory, has been found in suffering. Suffering, consciously undergone without retaliation, has been the fate, and the mark of self-identification, of the Jews during centuries. Thus the Jewish people, devoid of most of the paraphernalia of a self-continuing sociocultural entity, have in their turn maximized one aspect of the prime community, namely the uniqueness of the universe that *is* the prime community. Uniqueness seen, primarily, in its metaphysical, religious dimension. The chosen people. In diaspora, unorganized, but always ready to identify themselves and one another. To call this a nation, as is usually done, means stretching the word beyond the meaning that has been attached to it above.

Suddenly this "nation" converges upon a piece of territory – "Here we were born" – and begins to implement its nationhood in the fuller sense in which the word has hitherto been used in this paper. The first act is to strike roots in the soil: this having always been felt as being denied to the diaspora Jew (Baratz, 1954). But the first that the world realized about it was budding statehood: nowadays a *soi-disant* nation could not establish itself in a chosen area without appearing to be a state. In fact, even the Jews themselves had long been cherishing thoughs leading into exactly the same direction (Halpern, 1961). Thus, injecting themselves into the promised land the Jews turn out to be the state of Israel. In the process the land features as land to live in and as nothing else. There is realism in this, but it is of a compulsively visionary nature. Another order of realism, no less real for that, can but feature as accidental, and moreover as undesirable: the circumstance that this is not really empty land. The Jews have moved into Palestine much as Greek and later colonizers must have entered their colonies to be, only more so: that is, regardless of those already living there. They had and – tragic yet significant fact – still have no way of accounting for the presence of the Palestinian Arab. How much less could they, or can they, count with them; not to say, try to live with them. The voices of the few who would are drowned in the turmoil, not necessarily of contrary voices, but primarily of acute problems of survival, that is, of accommodating ever more immigrant Jews: the accumulative process of beginning self-consolidation and self-maintenance of the novel sociocultural entity.

14.3. In a summing-up that remains tragically inadequate to the actual implications of the problem yet renders justice to its essentials, one may say that Israel is an enclave in the Middle East in the last resort because as an attempt at nation building it is strictly a-typical. Whereas the Arab countries represent the normal features of an effort to overcome the problems of states that are not nations, Israel represents the converse effort to overcome the problem of a nation, of sorts, that is not a state. To this one must certainly add numerous vital differences in regard to all the ranges of determinants, conditioning as well as operational, that were listed above. Even so, it seems as if the problem may become less insoluble, the more headway either side to the conflict makes in resolving its own particular problem. Already there is no denying that the conflict as such has proven a great stimulus, on either side, to move ahead as fast as one can. The somewhat poignant question arises how beneficial cold wars might be . . .

15. The conclusion can be brief. The nation-state presents itself, most emphatically, as the natural unit (it should be better to say the only, the unique unit) of development: that is, of self-assertion and self-realization under thoroughly novel conditions – and this in developed and in underdeveloped countries alike. In the same manner the nation-state is tacitly accepted as the pre-established frame of reference, also the pre-established object, for all social sciences work taking place at the macro level.[16] But if in these two respects the nation-state with all its paraphernalia is a given thing, it surely is also a problematic thing.

[16] To offer just one, striking, example. F. Znaniecki's study on *Modern Nationalities* (1952) certainly represents as full a treatment of the phenomenon of the (Western) nation as may be expected from a sociology that is now about to be superseded by newer approaches. Even so, what it really can say is by no means in keeping with what it tries to do. The reason is simple but it is likely to go unnoticed. Certainly, the author seems not to have noticed it. It is that the entire work is based upon the tacit acceptance of the nation as a datum, and one of virtually universal(!) significance. What it fails to account for is that nations and nationalities are "situées et datées" before they could ever appear as perhaps more or less universal categories, whether of sociology or of societal life. An even more important matter is that it fails to account for the fact that the nation, over and above being a given thing, and prior thereto, is an ever elusive goal of social organization: and this simultaneously to being instrumental towards such organization.

We know this: in discussions on nation and nationalism more articulately, it seems, than in regular social sciences work. In a way, we are all the time reconsidering the true meaning of nation, nationality and nationalism, this being one way of carrying the realities thus named onward to some sort of fulfilment. On the other hand, this tends to render our commitment, even in attempts at reconsideration, such that we fail to achieve the mental distance needed for a clearer vision. Lack of clarity that, in its turn, could hardly be beneficial to our efforts, at national self-realization, at national development, at promoting international understanding.

ANNEX

In section 13.3 is was said that colonialism defeats itself. This is not merely an ad hoc statement. Its elaboration deserves a book, but there can be no harm in submitting here in briefest possible compass what it amounts to.

Formulated as a law, the statement made above should run more or less as follows. Any sociocultural entity, in fact any living institution or configuration will at a certain time need, for its continuing self-sustained operation, to rely on factors or elements that will eventually prove its undoing. Part of this law, it seems, has been found, and used ever since, by economists, namely in their law of diminishing returns.

But to see this law in its proper meaning one has to envisage it as one of two; and furthermore to envisage a complementarity relationship between the two. The countervailing law could, again in tentative formula, be proposed as follows. The emergence of any sociocultural entity, in fact of any living institution or configuration, is a matter of self-accumulation leading towards self-sustained existence.[17] It is again in economics that a glimpse of this law is discerned and put to use, namely in the somewhat odd-sounding notion of take-off into self-sustained growth.

The complementarity relationship that, from a logical viewpoint, is to be envisaged between the two propositions is, as usual, contradiction that does not end in a mutual cancelling out. Any sociocultural entity can thus be described as being determined by a mix of the two laws simultaneously, in which one prevails, ad hoc, over the other.

[17] Comp. my *Society as Process* (The Hague, Mouton, 1962), p. 88ff.

This view of the growth and decay of sociocultural phenomena offers a much desired improvement over the cyclical view, borrowed from early biology, that particularly the high priests of grand theory in history, and to an extent even in sociology, are wont to propound. The present view not merely makes it possible to envisage something like a cycle of growth and decay; in addition it allows fair presentation of abortive impulses, stunted growth, variety of directions into which growth may lead, sudden decay, slow decay and so many more phenomena that one encounters in studying sociocultural (and other) process.

The complementarity relationship, apart from having its logical aspect as its name suggests, also has an operational aspect. It could be suggested that operationally this complementarity appears as inherent selectivity.[18] Without selectivity, that is without a focus, the accumulation that signals growth, does not materialize. Without countervailing selectivity to cancel out any adverse implications of selective maximization of one element in the context of more or less comprehensive growth, self-sustenance is excluded, neither could decay become inevitable.

In the matter of inherent selectivity, it is usually asked "why" and "what". What is the pivotal element around which crystallization takes place? Why is it that it takes place at all, what puts it into motion? These questions come in endless quantity and they represent as many dead alleys for thought. The only viable question is "how". Or in other words, can one trace the determinants of the selective process, given selective focussing, in sociocultural process, upon one factor or element X and perhaps subsequently upon a countervailing factor or element Y? (Of course, in saying factor or element we need not necessarily refer to one simple entity: it could be complex in its own way.)

No doubt, views like these should not be important to the social scientist in isolation from other fields of investigation. In fact, as soon as he moves into this area of theorizing he will find himself covering ground that he has, to an extent, in common with some of the sciences on the one hand and with arts, history and philosophy on the other. Which could not render the subject less interesting.

[18] Koestler (1964, 209) employs the term selection in much the same way. Somehow selection must tie in with another phenomenon he discusses, namely scanning (p. 158f.). Now scanning is a known procedure in certain electronic devices, of which the mathematics and theoretical models are available. Theory formation on this matter in the social sciences is thus likely to benefit by what science has achieved in the connection, provided that in looking accross the fences between disciplines one bewares of the easy analogies.

REFERENCES AND SOME FURTHER READINGS

Akzin, Benjamin, *State and Nation* (London, Hutchinson, 1964).

Baratz, Joseph, *A Village by the Jordan: The Story of Degania* (London, Harvill, 1954).

Bennabi, Malik, *Vocation de l'Islam* (Paris, Seuil, 1954).

Berelson, B., and G. A. Steiner, *Human Behavior; An Inventory of Scientific Findings* (New York, Harcourt Brace, 1964).

Bidney, David, *Theoretical Anthropology* (New York, Columbia U.P., 1953).

Bloomfield, Leonard, *Language* (London, Allen Unwin, 1935). Reprint 1955.

Boulding, Kenneth, *The Image* (Ann Arbor, Univ. of Michigan Press, 1956).

Calvocoressi, Peter, *World Order and New States* (London, Chatto Windus, 1962).

Carr, Edward Hallett, *Nationalism and After* (London, Macmillan, 1954).

Cottam, Richard W., *Nationalism in Iran* (Pittsburgh U.P., 1964).

Deutsch, Karl W., *Nationalism and Social Communication* (New York, Wiley, 1953).

Doorn, J. A. A. van, en C. J. Lammers, *Moderne Sociologie. Systematiek en Analyse* (Utrecht, Spectrum, 1962).

Dunlap, Knight, *Religion, Its Functions in Social Life* (New York, McGraw-Hill, 1949).

Ellis, Willis D., *A Source Book of Gestalt Psychology* (London, Routledge Kegan Paul, 1955). First ed. 1938.

Ellul, Jaques, *La technique ou l'enjeu du siècle* (Paris, Colin, 1954).

Feibleman, James K., *The Institutions of Society* (London, Allen Unwin, 1960). First ed. 1956.

Forde, C. Daryll, *Habitat, Economy and Society, A Geographical Introduction to Ethnology* (London, Methuen, 1961). First ed. 1934.

Freyer, Hans, *Theorie des gegenwärtigen Zeitalters* (Stuttgart, Deutsche Verlags-Anstalt, 1963).

Geertz, Clifford (ed.), *Old Societies and New States, The Quest for Modernity in Asia and Africa* (New York, Free Press of Glencoe [MacMillan], 1963).

Green, Robert W. (ed.), *Protestantism and Capitalism: The Weber Thesis and Its Critics* (Boston, Heath, 1959).

Gross, Llewellyn (ed.), *Symposium on Sociological Theory* (Evanston, Ill., Row, 1959).

Grunebaum, G. E. von (ed.), *Unity and Variety in Muslim Civilization* (Chicago, U.P., 1955).

Haim, Sylvia G., *Arab Nationalism, An Anthology* (Berkeley, Univ. California, 1962).

Halpern, B., *The Idea of the Jewish State* (Cambridge, Mass., Harvard U.P., 1961).

Hertz, Frederick, *Nationality in History and Politics; A Psychology and Sociology of National Sentiment and Nationalism* (London, Routledge Kegan Paul, 1944).

Hoffet, Frédéric, *L'impérialisme protestant; Considérations sur le destin inégal des peuples protestants et catholiques dans le monde actuel* (Paris, Flammarion, 1948).

Huges, Carlton J. H., *The Historical Evolution of Modern Nationalism*, 4th pr. (New York, Macmillan, 1951).

—, *Nationalism: A Religion* (New York, Macmillan, 1960).

Jacobs, Norman, *The Origin of Modern Capitalism and Eastern Asia* (Hong Kong, U.P., 1958).

Jacobs, Philip E., c.s., *The Integration of Political Communities* (Philadelphia, J. B. Lippincott, 1964).

Kaplan, M. M., "The place of Religion in a Democracy", *The Review of Religion*, Columbia U.P., vol. 12, 1947-48, p. 179-192.

Kedouri, Elie, *Nationalism* (London, Hutchinson, 1960).

Kluckhorn, Clyde, *Mirror for Man; The Relation of Anthropology to Modern Life* (London, Harrap, 1950) (Paperback ed. Premier Books, Fawcett N.Y., 5th ed. 1963).

Knight, Frank H., and Thornton W. Merriam, *The Economic Order and Religion* (London, Routledge Kegan Paul, 1947).

Koestler, Arthur, *The Act of Creation* (London, Hutchinson, 1964).

Kohn, Hans, *The Idea of Nationalism; A Study of Its Origins and Background* (New York, Macmillan, 1946).

Leent, J. A. A. van, *De macro-micro-verhouding by de sociologie, de psychologie en de sociale psychologie* ('s-Gravenhage, 1962).

Lemberg, Eugen, *Nationalismus*, I: *Psychologie und Geschichte*; II: *Soziologie und politische Pädagogik* (München, Rowolt, 1964).

Lévi-Strauss, Cl., *Tristes tropiques* (Paris, Plon, 1955).

Madge, John, *The Origins of Scientific Sociology* (London, Tavistock, 1963).

Malinowski, Bronislaw, *A Scientific Theory of Culture and Other Essays* (Univ. N. Carolina Pr., 1944).

Manning, C. A. W., *The Nature of International Society* (London, Bell, 1962).

Marchal, André, *Systèmes et structures économiques* (Paris, P.U.F., 1959).

Nieuwenhuijze, C. A. O. van, *Cross-cultural Studies* (The Hague, Mouton, 1963).

Nisbet, Robert A., *Community and Power* (New York, Oxford U.P., Galaxy, 1962). First publ. 1953 as *The Quest for Community*.

Northrop, F. S. C. (ed.), *Ideological Differences and World Order; Studies in the Philosophy and Science of the World's Cultures* (New Haven, Yale U.P., 1949).

—, *Philosophical Anthropology and Practical Politics* (New York, Macmillan, 1960) Ch. 5: "What is a Nation?".

Nuseibeh, Hazem Zaki, *The Ideas of Arab Nationalism* (Ithica, Cornell U.P., 1956).

Parsons, Talcott, *The Social System* (London, Tavistock, 1952).

Pitt Rivers, J. (ed.), *Mediterranean Countrymen* (The Hague, Mouton, 1963).

Reid, Leslie, *The Sociology of Nature* (Harmondsworth, Penguin Books, 1962) (first ed. 1958 entitled *Earth's Company*, London, Murray).

Robinson, E. A. G. (ed.), *Economic Consequences of the Size of Nations* (New York, St. Martin, 1960).

Röling, Dr. B. V. A., *International Law in an Expanded World* (Amsterdam, Djambatan, 1960).

Schwarzenberger, Georg, *The Frontiers of International Law* (London, Stevens, 1962).

Samuelson, Kurt, *Religion and Economic Action*. Transl. E. G. French (Stockholm, Scandinavian University Books, 1961).

(Sapir, Edward), *Selected writings of Edward Sapir in Language, Culture and Personality*. Ed. D. G. Mandelbaum (Berkeley, California U.P., 1949).

Shafer, Boyd C., *Nationalism, Myth and Reality* (London, Gollancz, 1955).

Simmel, Georg, "The Number of Members as Determining the Sociological Form of the Group", *Amer. Journal of Sociology*, VIII (Chicago, 1962).

Slotkin, J. S., *Social Anthropology, the Science of Human Society and Culture* (New York, Macmillan, 1950).

Talmon, J. L., *The Origins of Totalitarian Democracy* (London, Secker Warburg, 1952).
Troeltsch, Ernst, *The Social Teaching of the Christian Churches.* Transl. O. Wyon (London, Alllen Unwin, 2 vols., 1931).
Wach, Joachim, *Sociology of Religion* (London, Routledge Kegan Paul, 1947).
Watt, W. Montgomery, *Islam and the Integration of Society* (London, Routledge Kegan Paul, 1961).
Weber, M., *The Protestant Ethic and the Spirit of Capitalism* (London, Allen Unwin, 1930).
Znaniecki, Florian, *Modern Nationalities, A Sociological Study* (Urbana, Illinois U.P., 1952).

II

LOCAL AND REGIONAL FACTORS IN NATIONAL DEVELOPMENT

Some Introductory Considerations

1. THE PRE-EXISTING SUBJECT MATTER OF THE SOCIAL SCIENCES

There exists a notable interconnection between, on the one hand, the state of affairs conditioning the emergence of the social sciences and, on the other hand, that which attracts the interest of those concerned with these sciences. Adam Smith has become the father of economics by writing a work on The Wealth of Nations. The point of gravitation of Hegel's social philosophy is the state as Hegel knew it. The same phenomenon can be traced with many other authors of roughly the same period. The thought of an entire culture period has been determined by a shift in the perception and appreciation of the political entities then serving as the frame of human existence. Political configurations are experienced as being significant in other respects besides the political. They are once again envisaged and experienced as contexts of human existence in a more comprehensive manner: political and also linguistic, cultural, economic. Society is once again experienced as togetherness, community, in principle of all those who are part of it. A significance is attributed to togetherness in human communities that is analogous to, be it not identical with, the significance attributed to the existence of man as an individual being.

This turning point in the perception of society constitutes at the same time the emergence of the social sciences – a form of intellectual concern about human togetherness which is manifestly different from inherited social philosophy. As generally recognized the main point of difference is the matter of preoccupation with things empirical. No longer it is considered satisfactory to establish the sense of human togetherness in speculative fashion: the intent is now to observe that sense empirically or even to establish it effectively. For present purposes the most interesting element in the emergence of the social sciences is that that which serves as the point of departure for the intellectual activities of social scientists is nothing but society as it happens to be manifest at the time and place concerned. No doubt this in itself is a matter of empiricism. Empiricism, however, that is bound to cause complications and thus to introduce a need for reconsideration, inasmuch as the social sciences constitute an effort to couple it to the urge-to-generalize which they have inherited from social philosophy.

2. THE QUERY CONCERNING THE OBJECT OF THE SOCIAL SCIENCES

It can thus be said that along with the emergence of the social sciences emerges the fundamental question concerning their object, their subject matter. Inevitably this question assumes the form of the need for reconsideration concerning the viability of the "vorgefundene", pre-established, object of these sciences – human society as framed in a given (nineteenth century Western) political structure. Certainly nobody will indulge in such reconsideration unless forced by circumstances. It may be occasioned by shifts in the political order; but also changes of an entirely different nature may promote it.

Now here is a fundamental question that will be answered differently according to the way which one conceives of the nature of the question itself. There is nothing exceptional in this and on this score alone the variety of possible answers is not a very interesting feature. This goes to the extent that even the effort to list the several answers could not be a rewarding enterprise for its own sake. The matter becomes important, however, if considered in connection with the ramification of the several social sciences: this in its turn reflects the ideas of social scientists concerning the nature and essence of the several branches of the social sciences.

Whatever the variety of answers given to the fundamental question just raised, they are bound to show the impact of the two main perspectives in which the question arises at the outset, viz., the empirical concern, not to say empiricism, on the one hand and the urge towards generalization on the other.

Between the two perspectives there does not exist a necessary logical relationship. Nevertheless it is tempting to perceive the simultaneous occurrence as a dilemma. This being so there is an ensuing temptation to restate either of the two perspectives in such a way that together the two will establish a logical relationship of opposition. One way is to present the empirical concern as a preoccupation with things specific or even unique that are of a fundamentally different order than things general: the result being two different orders of things between which an opposition can be construed. Alternatively the urge towards generalization may be restated as a preoccupation with universals that are manifest on a fundamentally different plane than that which is accessible to sense observation: with exactly the same consequence. One notices in passing that inherited oppositional constructs such as between things material and things spiritual are so close to what is under consideration here that the slightest occasion will suffice to cause these several issues to become confused.

The various ways in which the duality of perspectives just mentioned is handled in trying to answer the fundamental question concerning the object of the social sciences are bound to result in as many modalities of answering the question.

Furthermore the elaboration of each of these modalities of answering will be conditioned by accidental circumstances. Often the latter will attract so much attention for their own sake that one is inclined to consider them as the main determinants. Even so it will be attempted in the present paper to avoid separate discussion of these circumferential factors inasmuch as it would detract from that which must here be the primary concern, viz., to trace the main variants in answering the fundamental question concerning the object of the social sciences.

3. FOUR ANSWERS TO ONE FUNDAMENTAL QUESTION

In order to obtain a firm grasp of the object of the present paper it is useful to begin by attempting to gain some insight into the variety appearing in the answers that have in the course of time been given to the persistent question concerning the object of the social sciences. In fact, the subject matter of the present paper is but one way in which nowadays this same fundamental question happens to be raised. A discussion of the subject will therefore not be possible unless it accounts for the existing answers and their variety, and perhaps even opts for one from amongst them.

Where variety and multiplicity obtain, it is essential to avoid confusion. It will thus be useful to rely on schematic presentation. It is proposed to employ a four-fold scheme: the smallest number that will allow a reasonably articulate presentation, given the complexity of the subject matter at hand.

3.1. *First trend: empirical rather than generalizing*

Suppose somebody who is born and bred in the Western thought pattern. Two things in life have great attraction for him. On the one hand he will experience things as they are. On the other hand he is convinced that valid insight is necessarily general (capable of being shared by all and valid for all instances of the kind in question). His condition, as stated above, has many traits in common with the dilemma. Even so it is not a complete, inevitable dilemma. Accordingly he has an option between either acting as if there were a real dilemma (and thus implicitly aggravating his real condition), or acting upon the consideration that he is not (yet) facing a dilemma. In either case he has once more an option between two possibilities. First we consider what may be done upon the consideration that there is no dilemma (as yet). The way which is then open is the establishment of an order of priority. The option just referred to is an option between giving priority to either of the two elements in question at the expense of the other: it is a matter of either generalizing rather than being empirical, or being empirical rather than generalizing.

The first variant to be discussed is the procedure through which primarily the demands of empiricism are satisfied with a subsequent effort towards acceptable generalization.

The best example to illustrate this variant is that branch of the social sciences which nowadays is called social or cultural anthropology. (The two names represent differences that, however hotly debated, are of no concern to present purposes.) With reference to this example two observations are apposite.

First, the demands of empiricism have resulted in a reduction of scale. The anthropologist is accustomed to work far away from his own state and usually in regions where comparable political structures either do not occur or have been so recently introduced that for the small communities with which he is concerned they have little or no importance. Indeed his chosen field of activity is the small community of face-to-face relationships. His techniques of observation and analysis are entirely adjusted thereto. He meets the demands of empiricism to such a liberal extent that he may claim to have achieved a harmonious correspondence between object and technique of his studies.

Secondly. However, this good correspondence does not persist in regard to generalization. To many an anthropologist generalization is not merely second in the order of his activities, it is second in importance. It should be hard to deny that such generalization as one may encounter in anthropology could but exist by virtue of its utter unimportance. Were the average anthropologist to attach any importance to it, he would have panicked long ago. The existing state of affairs seems to be characterized by two points of view between which no mutual agreement obtains. On the one hand people decline to be concerned with any theory or theorizing; they will continuously underscore the particularisms of the one village they happen to have investigated. On the other hand there is the somehow ineffective urge towards, the more or less easy reliance upon, general theory or theory-in-the-making: which however is seldom explicitly applied as such. Embryonic theory or, alternatively, borrowed theory which, inasmuch as it remains up in the clouds, will obstruct clarification rather than promote it. Just how unsatisfactory this state of affairs is may become clear unexpectedly, for example when a hunting ground of anthropologists finds itself suddenly a component part of a larger political structure and when on that occasion expert advice is demanded from the anthropologist concerning the implication of measures concerned. It is on such occasions that specific knowledge of which the validity-because-of-generality (or of generalizability) is not manifest, threatens to end up in a somewhat precarious position. The concern thus engendered amongst a number of serious anthropologists speaks clearly from some of the writings of Prof. Lucy Mair (1957) concerning the application of anthropological insight for purposes of development policy. It is to be expected that this concern will lead to further consideration of the questions raised above.

3.2. *Second trend: generalizing rather than empirical*

The other variant shows the two elements in reverse order of priority: first generalization then empiricism. In this connection one may use certain aspects of economics by way of illustration, provided that one is ready to indulge in an even somewhat greater measure of simplification than was used in the example of anthropology.

In economic theory generalization has been offered full scope through the device of substitution. The notion state as the object of economic observation has been replaced by the notion scarcity: with the understanding that the latter can be understood as a shorthand indication, explicitly generalizing, of that which, in the state, attracts the economist's attention. Recall that the fathers of economics lived in societies in which the notion affluent society was unthinkable. No doubt, in this substitution of notions, which leads to the postulate that scarcity of means is the true object of economics, the national state is tacitly presupposed as the frame of reference within which that object is observed. Those in doubt on this score need but recall that entities larger than the nation are primarily considered by economists by means of a roundabout construct in which they prove the unexpected pupils of Grotius: they conceive of it as traffic between national economies. In its turn any entities of lesser order of magnitude than the nation-state, such as region, town, or village, are hardly ever observed as such: "for lack of statistical data for such entities separately". Besides macro-economics, economists will but consider micro-economics at the order of magnitude of the separate enterprise. This state of affairs is curiously illustrated by the term "an economy" which is in every respect the economic counterpart to the customary words country or state or nation.

Thus one finds in economics, as in anthropology, a shift of object in consequence of the option of an order of priority: in this case one that puts generalization in the first place and empiricism in the second. Starting out from the given political structure, one has proceeded to coin the notion of scarcity. Unlike what is the case in anthropology, the substituting notion in economics is clearly of a different order than the substituted one; on the other hand the substituted one, that is the original indication of object, does not entirely disappear but continues to play a role in the background.

Thus economics satisfies the primary demand – in this case of generalizability – in a manner that differs from, but is not in itself less satisfactory than, the manner in which anthropology meets its primary demand, viz. of empiricism.

The question is now whether economics is more successful than anthropology in the matter of subsequently meeting also the secondary demand: which in the case of economics is empiricism. At this point effective comparison becomes difficult, more so inasmuch as empiricism features in economic thought in more than one way.

In the first place the generalizing-abstracting thought of certain schools of economists has built a bridge towards empiricism by means of constructs based upon mathematics and upon quantification of factual data. In this way it places itself in more or less open communication with empirical reality, be it that the latter remains represented in one very specific fashion.

Furthermore economists of various schools have a strong policy inclination. To them economic theory derives its value, its significance, from application for policy purposes. It has an inherent tendency to seek creative confrontation with empirical reality.

The third point that must be listed in the connection shows the reverse side of the medal of which the two previous points showed the bright side. It is in the context of policy-minded economics that the notion "development" has initially been determined. In this notion however, as is gradually becoming clear, one can not merely trace the positive possibilities which are at the disposal of economic thought in regard to empirical reality, but perhaps rather its limitations. In the hard school of practice, development proves to embrace fundamentally more than economics can ever be counted to cover, given the manner in which it has grown and become implemented: starting out from the definition of its object as just mentioned. In the light of development problems such as we know them nowadays from "most of the world", the economic notion of development – or should one say, the notion of economic development – is inadequate. In stating the fact one does not do the slightest injustice to economics; nor, for that matter, to economists. Between definition of object and urge towards policy, economic thought has a fundamental lack of clarity. The source of this lack of clarity can but be the insufficient extent to which theorizing meets the secondary demand, viz. empiricism, upon having more (not to say, completely) successfully met the former demand, viz. generalizability. No doubt the manner in which this former demand has been met will also play its role. Once more the conclusion is that the secondary demand leads to considerable difficulty and that again it is the matter of development problems which has brought these difficulties into the open. Here again the attention is called back to the fundamental question of the determination of the object of the science in question. Nothing proves this more clearly than the circumstance that in economic development policy anywhere in the world the oft repeated but seldom answered question refers to the "natural unit of development".

3.3. *Third trend: the search for a compromise*

The other way to conceive of the simultaneous occurrence of the demands for empiricism and for generalization is to construct them as a full-fledged dilemma. Starting out from this second construct – which is likely to occur tacitly rather than explicitly in most cases – one has once more an option between two variants. Characteristic to both variants is the consideration that the dilemma cannot be resolved by elimination of one demand to the benefit of the other. Thus the one variant represents the attempt to find a way out of the dilemma by means of a compromise; the other tries to achieve the same result by means of a "tertium datur": that is, by postulating a third demand, which, by rendering the two original demands superfluous, appears to eliminate the problematic opposition that obtains between the two without having to actually resolve it.

By way of illustration of the former variant of this position, the compromise, certain traits of sociological thought may be referred to: surely at the risk of offering a presentation of the discipline concerned that may, to an extent, sound like a caricature. Sociology has been launched upon its career with the claim that it was the universal science of society. Its name, and not merely its name, implicitly maintains this claim up to the present day. Thus, sociology represented the attempt to perform what economics had ceased to perform on account of its focussing upon a more or less restrictive object, as discussed above. This places sociology in the position of becoming a somewhat belated heir to the pre-established object of the social sciences as discussed above: a heir that has not even managed to show a consistent awareness as to its heritage. A less direct, but for that reason not less clear symptom of the universality claims of sociology can be traced in the systematic adoption and maintenance of a terminology, meant for scientific use, which is derived from common speech – a kind of uncleanliness the elimination of which goes slowly and with much resistance.

In the field of tension between empiricism and urge towards generalization sociology shows one and the other of the shifts discussed above in regard to the definition of its object. Occasionally the notion indicating the comprehensive society as embraced by one's own historically given political structure is replaced by the notion group. Thus employed, group does not merely feature as a constituent part of society at large; sometimes it will also feature as a "pars pro toto". In either meaning the secondary notion represents a restriction of scope as compared to the primary notion: a restriction of the object of the discipline concerned comparable to the one described above in connection with anthropology. On other occasions one encounters a recasting of again the same primary notion much in the manner discussed above in connection with economics. In such cases conceptualization becomes focussed upon one particular qualitative aspect of the entity from which one had started out. Thus, as one example out of many, the notion of conflict.

It is no doubt possible to envisage a certain measure of interaction between the two tendencies just signalled. Upon further consideration it should even be possible to express the expectation that this interaction in its turn bestows upon sociological thought a certain sense of direction – traceable though not overly manifest – in regard to the question what is the object of sociology. It looks as if in sociology we are moving towards a conception of its object which is entirely ad hoc, in the sense that it is a function of the working hypothesis from which one happens to start out in a given case. Mutual determination, in other words, between object of observation on the one hand and theory on the other. Under such conditions one can not say much more, a priori, on the object of the discipline concerned but that as a rule it will be neither extremely comprehensive (such as total society) nor will it be extremely limited (such as for instance the family). This is in keeping with middle range theory as fashionable nowadays. What middle range theory does not necessarily indicate is that the scope or size is in no way decisive for the manner in which the object of the discipline will be determined. The decisive role in this regard will be played by other categories of a more operational nature. In this sense one could perhaps say that a social unit, a sociocultural entity, is sociologically relevant inasmuch as it establishes an instance, an ad oculos demonstration, of a sociological working hypothesis.

Much of this development belongs to the future. In the practice of application of theoretical insight, particularly in development work, sociology is still too often found in the role of Cinderella – with, amazingly, economics in the role of the beautiful sisters. The reasons are only too clear: a burdening heritage of useless concepts and a fundamental lack of clarity, still lingering on in regard to the simultaneous demands of empiricism and generalization.

3.4. *Fourth trend: evading the dilemma and search for a "tertium datur"*

One may also try to evade the dilemma, viz. by making the two demands disappear behind one overruling demand. Where this happens the opposition between the two loses its significance. The problem remains unsolved but its validity would have disappeared.

Two examples will help to clarify the position. This time no branches of the social sciences will feature as examples. Instead we shall refer to a tradition of thought on the one hand and a particular working technique on the other, each of which leads a more or less independent existence and can be traced in more than one discipline.

The tradition of thought that is exemplary for the position under consideration was already mentioned in passing, namely middle range theory or, with its older name, principia or axiomata media. To the followers of this tradition universals on the one hand and particulars on the other establish a Scylla and Charybdis between which one has carefully to steer one's course. Their objection against a good deal of inherited thought, both of social philosophy and of social sciences, is that the jump from specific observed fact to general notion cannot really be made. To some of them this is a problem of an extravagant difference in order of magnitude, to others it is a difference of logical order (the traditional problem of induction versus deduction is likely to play a role in the connection). Characteristic for this tradition of thought is Karl Mannheim, whose purpose is action (planning for freedom) but who wants to act in a scientific manner: that is, upon the basis of general insight. His principia media are on the one hand conceptual tools for the planner, on the other hand they are configurations of the proper scope to observe comprehensive social change. With this heritage of Bacon, Mill and Mannheim, Robert Merton has managed to launch a sociological fashion. But it is particularly Zetterberg who has with a beginning of success attacked the fundamental logical weaknesses inherent in the construct of the middle range. Upon closer inspection one finds that he has been able to succeed in this regard by leaving out of consideration the extreme positions, which in fact constituted the real impulse leading to the construct of the middle range. To him, empiricism features as verifiability; generalizability in its turn features as relative inclusiveness. Thereby the tradition of thought that concerns itself with the middle range has ended up in exactly the same tendency as a good deal of other modern sociology. What is interesting in the connection is that Merton's fashion of the middle range operates mainly as a kind of smoke screen behind which can take place a fundamental reorientation that should have encountered considerably more resistance without such unintended protection.

The other example is the technique called "area study". It is a child of the Second World War; the primary concern of its instigators was not necessarily scientific. Its purpose was to accumulate knowledge concerning actual and potential enemies and, one assumes, also friends. This technique has survived the war and has, primarily in the USA, found a place within the framework of attempts to find one's bearings in the "one world" of the post-war period. In the connection there has been no shortage of debates on the definition of area. Eventually, under the influence of anthropologists like Ralph Beals, the debate concerning the ideal scope of the area has subsided and made room for the awareness that area and working hypothesis determine one another mutually. This awareness in its turn leads once again to the conception of the object of the social sciences already signalled above: a pluralist conception entirely in terms of reciprocity between working hypothesis and its actual manifestation in a sociocultural entity determined ad hoc.

4. COUNTERVAILING TRENDS

There are thus at least three impulses that one may interpret as leading into the same direction. It should on the other hand be unrealistic to expect that presently these will merge in such a way as to eliminate the pre-established object of the social sciences, that cumbersome heritage of an undesired past, and to replace it by a scientifically unassailable conception.

The world constellation of today, with its cold war and its development problems, is said to be dominated by neo-colonialism and neo-imperialism; in fact it stands in the sign of neo-nationalism. Until World War II, the world was composed according to a two-fold pattern. On the one hand there was a network of political and ensuing economic and other relationships, friendly or antagonistic, multilateral and bilateral, between political structures rated, and mutually rating one another, as respectable, as up to par. Complementary thereto, specific relations existed between some of these states and certain territories in the world where no such political structures occurred: all this in such a manner that the latter territories featured in the comprehensive world picture as a kind of second plane. In the former regard, World War II has resulted in fundamental shifts; in the latter regard it has caused a complete reversal of conditions. In consequence of these two together, the world faces the vital problem of being in need of a totaly new constellation: for which no examples are available. This problem has as many names as it has manifestations. Much confusion results from the circumstance that practically none of these indicate it in its full width and depth.

A central feature is the ubiquitous need for new integration of fundamentally all components of humanity: in such a way that they will effectively, not to say equally, share in the world constellation to be. This need manifests itself in different ways according to whether a given political structure is part of the former primary plane of the (pre-war) world constellation, or part of the former secondary plane. In regard to those components of contemporary mankind belonging to the first-named category, one can even further specify the manifestations of this need. Between the big ones there is a struggle for hegemony; between the smaller and smallest ones there is a grim struggle for self-maintenance; and amongst all there is an urge to effectively participate in the effort to give shape to the new world constellation: by means of action in regard to one another as much as by means of action in regard to political entities belonging to the former secondary plane. Components of contemporary mankind belonging to what used to be in pre-war days the secondary plane of world organization, show the same need in their own way. Often it is an overheated need for self-realization: and this within frames of reference that in their turn remain to be implemented in and by the same process of self-realization. The new nation-state, the demand for social justice, the need for development are equally symptomatic for this urge: the urge, the desire to perceive one's own social identity, whichever that may be, clearly manifest in the present world constellation. Clearly, (under)development and nationalism are two names, from amongst many, for one and the same thing. The thing in question is the fundamental query concerning the natural unit which is capable of being a component to the new world constellation to be achieved.

Upon closer consideration this query proves an open question. However, to an excited first inspection it may give the impression of having found its proper answer long ago. (There is a parallelism between this situation and the situation prevailing at the time of the emergence of the social sciences.) No doubt, the answer then suggested refers to the pre-existing political structures of the former primary plane of world organization: political structures that, it is then suggested, would have to be reproduced for the benefit of the political entities that have now come into existence at what previously was the secondary plane of international organization: and this in lock, stock and barrel fashion, or at least regardless of the crisis which the same political structures are now passing through in the countries of their origin. It is most unlikely that such a formula could work. However, before one is able to recognize this (if ever there comes a moment at which one could afford to recognize it), much confusion and possibly a great deal of evil have already been brought about. To repeat, all this is neo-nationalism: a dream that is slow in coming true, a fundamental social malaise. From this brief survey it follows that amongst the existing components of mankind, both those belonging to the former primary plan and those belonging to the former secondary plan, no real answer is available to the burning query concerning nature and scope of the *natural* component of the world constellation of tomorrow, which actually should have been in existence already today. We know exactly who nowadays are members (or, for that matter, are not members) of the United Nations; but this knowledge causes grave concern about the question who actually should be members of this kind of organization.

Yet another manifestation of always the same comprehensive order of problems is that which is currently called development problems. It is somehow more noticed in regard to entities belonging to the former secondary plan than in regard to those belonging to the former primary plan. There is however a growing insight that between the two there is no fundamental difference. Underdevelopment is a typical phenomenon of the present period. Once again it entails the query about the natural unit, this time unit of development. The answer that tends to be given to this question is determined, amongst other things, by the fact that the notion of development, inasmuch as it is a partial indication of the comprehensive problems concerned, is synonymous to such indications as the establishment of new states, independence, and the like. In other words, it is likely that people will indicate as the natural unit for development that which they will also indicate in connection with other aspects of the problem of social self-realization: that is, a political structure, a state analogous to the Western political structure of the nineteenth century. Much as it seems natural to do this, there is no way of envisaging the consequences. It is simply an emergency jump into the dark. There is an additional incentive to jump, in the circumstance that development is often presented as being primarily, if not exclusively, a matter of economic development. This lends a measure of preponderance to economic reasoning: with its traditional frame of reference, viz. the state.

There is thus a tendency to rely on the state, a political structure postulated ad hoc and in an entirely uncritical fashion, as the only natural unit for a wide range of vital activities. From a certain viewpoint, this tendency cannot evade being branded reactionary. Development work, being one of the activities in question, reflects a trend of reasoning in the social sciences that shows less and less willingness to consider one historically given type of sociocultural entity as *the* one and only natural operational unit. The same tendency considered from yet another angle is likely to be called purely tentative not to say experimental: the political structure that has to serve as the frame of reference to ever so many measures is in its turn a plank of the platform.

5. UNITS OF LESSER ORDER OF MAGNITUDE

The importance of the countervailing tendencies discussed in the previous section is not such that they have prevented people from considering, and recognizing the importance of, socio-economic entities of lesser order of magnitude than the state: and this particularly for purposes like development.

There has been a persistent interest in units determined otherwise than in terms of the political structure of the state. One example is a field of studies and activities commonly indicated by the adjective regional: regional strategy, regional organization, regional planning and the like. In the connection, the region tends to feature as a unit which for given purposes is more natural and therefore offers more and better chances than does the state. (Implicitly this contains a judgment concerning the validity of historical-political factors as determinants of socio-economic units.) Occasionally such a region may embrace more than one state. Then it is likely that the component states will play a quite noticeable role in their capacity as components, regardless of the question whether their characteristics entitle them to such roles given the purpose concerned.

On other occasions the chosen region may be smaller than the political expanse of a given state. This latter state of affairs occurs in many regional-level types of organization and similarly in a field of applied scholarly activity commonly called regional planning. One notes in this regard that current discussions tend to concentrate upon matters of integration, in the sense of integrated plurality of aspects, integration in the application of various branches of science, social as well as natural. Less attention is devoted to the problem of integrating the chosen region into the encompassing entity of which it is a component part: entity that is often but not necessarily determined according to the political criteria of statehood.

Another example of active interest in lesser – in this case indeed small – units can be found in community development and in adult education. The two are different in many respects yet comparable for present purposes. Here again the tendency is to reject the state: as being too large, as being insufficiently a real unit of living together, if measured by the yardstick of specific purposes which in their turn usually are variants on the theme of development. This, and in addition the conscious option for the geographically smallest social entity: as a rule the village or comparable urban units. Within a framework like this there is less scope for impersonal, more or less scientific policy and accordingly more scope for effective goodwill. This has advantages as well as disadvantages. Quite a few do-gooders who feel attracted by the ideology of development – whatever that may be – find in this kind of work a field of activity that is often innocent rather than effective. Effective, then, in the sense that the result of their work would stay after their departure or that changes induced by them into the community where they work would be spontaneously adopted by surrounding communities. It is beyond doubt that the comprehensive society lives primarily in the small unities. However, this verity has little to teach concerning the consequences that changes "sold" to the small community could or should have for society at large (which, as stated above, is all too often conceived as being determined politically).

In the same connection one may point to something which could not serve as an example of activities in a framework other than the state but that nevertheless has the same importance as the two examples listed, given the query concerning the natural socio-economic unit (for a given purpose, such as development according to one definition or another). This is the circumstance already mentioned that in many of the recently independent states the comprehensive political framework of the state means less, and is considerably less effective, than in the older states. The reason is that socio-economic units, of various natures and scopes, have merged into the comprehensive political units called states, and this with a variable degree of consciousness and willingness so far as these units were concerned. In the process the original entities are likely to have been profoundly shaken up. Even so they retain for the time being a greater measure of real importance and effectiveness than has the new comprehensive unit, the nation-state. Considered in the same light, the latter remains too incidental, too ineffectively part of the pattern of orientation and of expectation of those who may (or must) consider themselves its citizens. In one way or another a development policy, in order to be realistic, has to account for this kind of circumstances. Accordingly it is likely to account for, not to say to rely on, the older, more effective components rather more than would be the case under different circumstances. This is not to say that the older components of these new states offer a ready and easy solution to the problem at hand. Often the new state is unlikely to persist without fundamental changes in, or even liquidation of, the pre-existing units.

Situations like these are as many demonstrations of the problem of offering an effective framing to that which is done in regard to component units, within the context of the comprehensive unit concerned.

In yet another manifestation, the same problem can be phrased as a problem of incidental versus consistent approach in the matter of what can be done in regard to the various component entities taken together. What does a given regional plan mean for the adjacent region? Does a village project effectively function as an example for neighbouring villages?

Again the same problem may also be phrased from the point of view of the comprehensive entity, that is as a rule from the point of view of the nation-state. To this viewpoint it appears as the problem of effective interaction – some would speak of reciprocity – between the so-called centre, the capital, the point where the decisions are taken, and on the other hand the periphery, the countryside, the province, where decisions are to result in more or less the same effect although implemented necessarily in mutatis mutandis fashion.

In yet another variant of phrasing one may put the same problem as a problem of integration, of promoting an equally high degree of participation, by all component units, in the common wealth.

The solution of this problem must take place in the face of a wide range of difficulties. Most of these originate from one out of two sources. The one is the circumstance that one cannot start from a clean slate and that on the other hand most of the inherited structures, interests and orientations are undesirable ballast. The other is that even if one could start from a clean slate, one would have no example available to show that which one has to do. The consequence is that one has to make do with whatever one finds present: an excessive concentration of economic, social and cultural potential in the capital, privileged elites (with old or perhaps newly established people composing them), lack of communication between the component parts of the country, and many more similar features, all of which have a strong tendency towards self-maintenance and even towards self-reinforcement: and this regardless of the fact that nothing is so necessary as to replace them by a new state of affairs. In short, a condition in which the very concept of revolution loses its sense and its power. It is good to state this in so many words, inasmuch as precisely on this point there exists a great deal of misunderstanding. All too often and all too easily it is said that a revolution is actually going on, or that a revolution is necessary, or even that a revolution will be made if such and such demands are not immediately met. Measured by the standards of what is actually happening, this is idle talk. The comprehensive complex of shifts and changes in question is of a different order of magnitude than revolutions are. Anyone who under these circumstances speaks of revolution, preaches revolution, or hopes for revolution, is either uninformed or in bad faith.

6. PROVISIONAL CONCLUSION

The preceding considerations may be summed up in three points.

(1) "Development" is a summary and inadequate indication of a comprehensive process of change that takes place in the entire world but that does not have the entire world as its primary frame of reference. Its primary frame of reference is the larger socio-economic unit which is commonly indicated by the name state: a name that as such is heterogeneous to matters of development. It is an indication by sheer coincidence: the label does not fit the bag but it was readily at hand. Secondary frames of reference for development may be either larger or smaller. On the one hand there are the more comprehensive geographical, economic and geopolitical units, seldom effectively institutionalized as such, and also mankind at large. On the other hand there are various kinds of components to the national economy, which is also called the national community or the state. In the latter, the economist may distinguish sectors or, on an even lesser scale, enterprises or groups of enterprises; the sociologist in his turn will distinguish groups of varying nature and size and, on the smallest scale, families and primary groups, interest groups and the like. It must be said that thus far none of them postulate a clear pattern of interconnections by virtue of which all these components would establish the comprehensive unit. Their partial notions are partial abstractions: not components of comprehensive abstractions.

Summing up, the natural unit for development receives its name as a matter of coincidence; accordingly the name is usually inadequate. More in particular, it proves very difficult to account for the fact that the entity thus named is not merely an entity or unit per se, but at the same time is both a constituent element to a more embracing entity and a complex phenomenon of which the internal cohesion remains problematic.

(2) Development situations are characterized by an excessive occurrence of factors that hamper development. These factors will specifically block the integrative development of the comprehensive socio-economic unit in question: by hampering the achievement of a balance between component entities. The capital will live at the expense of the rest of the country; the elite will live at the expense of the masses; monocultures will be maintained at the expense of other cultures. Too many people and groups of people do not get what they should get nor do they give what they should give: economically, socially, culturally. Even in cases where the nation-state is sufficiently established to offer a foothold for measures to be taken, it is seldom clear where one should begin.

(3) In the social sciences there is a tendency to advance a more and more flexible answer to the query concerning the natural sociocultural unit for purposes of observation or of action. The average economist is perhaps slightly more conservative than some of the other disciplines in maintaining the conception, gradually becoming obsolete, that the state, the national economy, is the (only) natural frame of reference for social sciences thought. The gradually prevailing trend seems to lead into the direction of ad hoc determination of the entity that one chooses to consider: and this then in the light of a given theoretical interest (which in its turn could well tie in with a policy interest). A fundamental orientation for which such notions as pluralism, relativism and operational effectiveness would seem to be major determinants.

7. LOCAL AND REGIONAL FACTORS OF NATIONAL DEVELOPMENT

Thus in matters of development it is likely that one will in the first place think in terms of the nation-state. This is not so because the nation-state, as a primarily politically determined unit, would be the only and entirely adequate frame of reference for matters of development. Rather it is a matter of neo-nationalism, as symptomatic for the comprehensive change which in its turn is also described as under-development, need for development and the like: all this regardless of the question concerning the appropriateness or usefulness of the national political structure, derived from the nineteenth century Western world, as the institutional frame for the socio-economic units in question.

The social scientist can afford to accept this state of affairs, be it with certain reservations both of an empirical and of a fundamental nature. Truly the political structure of the nation-state is no longer the sole frame of reference that he recognizes. As stated above, the trend in this regard is much to the contrary. On the other hand, within the plurality of viable frames of reference the nation-state is by no means excluded, so long as it can be conceived in relationship with an adequately phrased problem: a relevant theoretical proposition. The usefulness of this, as of any, frame of reference depends entirely upon the further elaboration of the definition of development.

On the other hand the same considerations inspire the thought that there is no reason whatsoever to conceive of development exclusively in terms of the national state. Much rather does it seem advisable to avoid such one-sidedness and to consider, besides the nation-state, also other, both larger and smaller, frames of reference. Already one finds a considerable number of agencies that work in an increasing measure according to this insight. The tendency, in so far as international agencies are involved, is to develop regional activities on a scale larger rather than smaller than the national state. On the other hand national development agencies, as well as international ones co-operating with these, are increasingly concerned with lesser "units of development".

In the latter case the question is which entities are to be considered. It is only too natural that one will try to turn necessity into virtue by attempts to operate pre-established entities much in the same way as the emerging social sciences have done in their turn in regard to the nation-state. Thus, existing administrative or geographical regions will claim attention and, on a lesser scale, such units as towns and villages. Once again the difficulty is that none of these entities is primarily conceived in terms of development. The question is therefore whether they can appropriately be presented and treated as units of development. However, what has been stated above on this kind of questions should suffice again in the present connection. Even if one feels obliged to maintain certain reservations, the attempt remains worth the effort. The point is that practice indicates the desirability, if not necessity, of development work that, whilst taking place within the national framework, is to be implemented at the level of the several components of the national complex. This being so, there is little alternative but to try and make the best of existing units.

Let it be repeated that to closer critical inspection many problems are left, exactly as in the case of the emerging social sciences employing the state as their pre-established object. These problems must be considered in due course of time and the recognition of their existence is no sufficient argument for not trying to make the best of existing units. To see the importance of this consideration it suffices to consider any alternatives that would be available, were one to refuse the entities just mentioned for the purpose at hand. The one alternative would be to try and meet the demand in question by means of the units summed up above, the ones traditionally distinguished by the several social sciences. In so doing one would encounter a preliminary problem that thus far nobody has been able to solve, namely how to homologize all the diverse kinds of entities distinguished by the several social sciences. The other alternative would be to bank upon future developments in the social sciences and to look for entities determined in the ad hoc manner briefly discussed above. The risk in this case is obviously that it could be very long before the developments in question would have advanced to such an extent as to make the minimum number of conceptual entities available for use.

8. CONCLUSIONS

There are one or two conclusive observations to be made that may serve to round off the present orientation.

The ultimate question is not whether in a given town or area a certain type of development can successfully be instigated. This in itself can be difficult enough; but the crucial point is what such development means for, and in the framework of, the encompassing – let us say, national – context.

In the last resort, this is a matter of mutuality of perspectives. What does local or regional development, whether potential or actual, mean within the comprehensive national framework, considered from the viewpoint of the "centre"? Conversely, what does the national framework mean for local or regional development that has to fit therein – in other words, that has to contribute to the comprehensive whole and that has in its turn to receive impulses from this comprehensive whole? A complicating consideration is that when one speaks of the comprehensive whole one may actually refer to different things. It may mean the centre, it may mean any incidental combination of other local and/

or regional entities, and at again other occasions it may mean the national state in its totality. Whatever the meaning, the mutuality of perspective will have to be accounted for in manifest forms of interaction.

An additional consideration is that this is in fact a need for the establishment of operational balances between elements that are in principle non-congruent, of different order and often non-equivalent. "Balanced economic growth" is not so much a matter of equal chances as a matter of an optimally favourable distribution of entirely diverse chances. In other words, it is a kind of division of labour: economic, but at the same time social, cultural, political and administrative. A matter, besides, of striking the balance between qualitative differences that in their turn ensue not merely from geographical location in regard to means of communication and raw materials, but just as much from the order of magnitude of the socio-economic entities in question. In the connection one cannot suffice by establishing orders of priorities. What is at stake in principle is the division and adjudication of unequal chances of many, highly variable kinds: in such a manner that the cumulative effect of all these chances, if and when realized, should be optimal. Surely chances of which people avail themselves are more important than chances that are merely offered. With this simple proposition the accent in the total distributive mechanism shifts once again – away from the calculations of centrally located planners and towards the less controlable, more spontaneous action in the framework of local and regional entities: severally and once again as components of the comprehensive national whole. Which once more raises the question of balance.

There was once a time when people knew the answers to these questions: "prestabilirte Harmonie", pre-established harmony. The present generation has lost the deistic faith from which this answer sprang. What is left to them is a persistent engagement, the relentless necessity to apply general insight according to best knowledge and competence in incidentally given cases wherever they find work to do, whether of a theorizing or of a policy nature. Every single local or regional entity represents a case of this kind: and this neither more nor less so than the national entity.

In order to work effectively towards development in a local or regional context one should in particular avoid two errors. First, to think that it is unnecessary to look beyond the local or regional entity in which and for which one works: the encompassing entity is important for more reasons than that one may refer to it for support. Secondly, to think that one should be able to work effectively without accounting for general theoretical insight. Conversely, from the viewpoint of any central development agency apparatus, the local or regional entity is not merely a "grass-roots level" with which one tries to maintain contact, in one way or another, for the purpose of obtaining factual data and feed-back towards the execution of plans inspired upon theoretical models (or upon the wishes of the bearers of political power). Much rather it is a real factor of national development: a factor that demands to be recognized, evaluated and accounted for, and that furthermore should play its own role in its own right.

ANNEX: SUGGESTIONS FOR RESEARCH [1]

We wish to investigate the importance – whether positive or perhaps negative, whether as an asset or as a liability – of lesser entities within the national framework, given the circumstance that, rightly or wrongly, the latter tends nowadays to feature as the main, not to say the natural 'unit of development'. These entities may be economically determined, or perhaps sociologically, or politically or administratively: in one or more of these ways, they will lead a life of their own, have a specificity of their own, represent possibilities or limitations in their own right, and as such they will tend to constitute distinct factors in the national development context. Factors that are actually taken into account or perhaps factors neglected; factors likely to shape up as drawbacks or perhaps promotive factors.

In focussing attention on this order of phenomena we hope to produce facts, knowledge and insight which, once brought to the attention of those actively responsible for development planning and action, will prove helpful towards more and more widely, perhaps also more evenly, effective development measures.

In thus presenting the theme one avoids the suggestion that national development should (or should not) be balanced in the sense that an even distribution of developmental impulses should be achieved over any local and/or regional components of the nation state. Maybe it should, maybe it should not. This kind of thing, along with many similar

[1] Abbreviated version of the report of a consultative seminar of the Mediterranean Social Sciences Research Council, held April 8-10, 1963, at the National Hearth Centre of the Royal National Foundation, Athens.

postulates that need not be listed here, remains to be seen: after the investigation has been brought to a conclusion.

In elaborating the theme, one distinguishes rather easily between two main aspects or perspectives, each of which then calls for further specification. One is the *setting*; the other is the *operational aspects* of the problem.

1. In regard to the *setting* in which the problem of "local and regional factors in national development" occurs, one may distinguish between *three main determinants*. Each of these determinants constitutes a valid heading under which more specific research activities could be undertaken. They can also constitute valid themes as such, namely when studied for a particular local or regional entity exclusively.

a. The first determinant is the *prevailing conception of the underdevelopment situation,* for purposes of development planning and action, according to one out of a number of existing models, such as,

(1) the model of "pockets" of development in underdeveloped surroundings or, alternatively, of underdevelopment in developed surroundings.
(2) the model of a "dualist" economy and/or society, in the sense of J. H. Boeke.
(3) the model of a "pluralist" economy and/or society, whether (a) according to developmental stages, or (b) according to socio-economic, perhaps also cultural segmentation.
(4) the model of "discontinuity", which differs from the previous models particularly in abstaining from any a priori determination as to the precise nature of the non-integration that happens to affect development in a given case.

The practical importance of investigations concerning the often preconceived and occasionally also tacitly presupposed models that occur in given cases is evident. More often than not goal setting in development can be clarified when the presupposed model concerning the prevailing to-be-developed situation can be made explicit. More so since it is likely that one will find that several models play a role at once or at different moments all in one context, or that a model prevailing at the local level may differ from the one prevailing at the central level of one and the same development project. In short, investigations on this matter stand to bring about some usually unsolicited clarifications of a most fundamental nature, which stand to prove most helpful and influential for further development practice.

b. The second determinant can be found by *establishing* the *types* or *patterns* of *actually prevailing disparities.* In contradistinction from the first determinant, which runs in terms of conceptualization on the part of those concerned in a given case, the present determinant is conceived in terms of concrete social, economic, administrative, political and-what-not discrepancies actually prevailing in a given case, say, of national development (or of development in a given part of a state): discrepancies that are so many signal phenomena bringing out the importance of considering local and/or regional factors in national development as factors per se.

Again, the practical use of investigations along these lines is to be seen mainly as a matter of fundamental (in this case also very factual) clarification. Much less important is an angle that may in certain cases tend to appear more conspicious than that of fundamental clarification, viz. the possibility that actual discrepancies might be signalled that would demand to be explained (or perhaps even remedied) as such, regardless of their significance for the more profound problem, what is the role or importance of given local and/or regional factors in a context of national development. It may be advisable to avoid the inevitably side-tracking effect of this additional aspect as much as possible, and to concentrate research concerning prevailing discrepancies upon the effort at bringing about a typology of discrepancies. To have such a typology is highly important. It should constitute one of the main means to assess the real role and importance of particular local (or regional) entities in a context of national development, whether as (actual or potential) assets or as (actual or potential) liabilities.

c. The third determinant is the *prevailing image of development* (or – more likely – *range of images* of development) in the given national context: image (or range of images) that is, no doubt, necessarily *culture-conditioned* by and in this context. This range is necessarily very wide, inasmuch as it runs from highly sophisticated concepts (of advanced economic and other social sciences theory and of subtle social philosophy of many schools), all the way through the simple and entirely limited dreams of betterment that may be cherished by the man in the shanty town. In addition, there is likely to be plurality of images, not just when several or many minds are to converge upon concrete action, but even in one mind. Similarly, the image-mix, the various images simultaneously occurring in one mind or group of minds, is likely to vary with place and time. Clearly, this matter is as vital, from a viewpoint, once more, of fundamental clarification, as it is complex. Investigation may have to set out from the choice of a rather narrowly circumscribed geographical scope, which should at the same time account for the matter of cultural conditioning. Then, actual investigations may have to start out from a provisionally adopted (if necessary a priori) typology of images of development. (The typology is then, of course, subject to subsequent verification of its validity, by applying it.)

2. In regard to the *operational aspects*, two complementary viewpoints must be distinguished, each offering ample scope for investigations. Besides, the interaction between the two viewpoints demands attention and offers in its turn scope fur further research projects.

a. First, the viewpoint of *integral development of the entire nation:* social, economic, administrative, political. The view moves from the encompassing national entity towards the component local and/or regional entities. What one will thus seek to establish is the degree of "compliance" on the part of the latter that one seems entitled to expect – or alternatively that actually manifests itself – in regard to development planning and action of nation-wide scope and thus necesssarily originating at the "central" level. More specifically, one will want to establish the degree to which and the manner in which the various local and/or regional entities that one chooses to distinguish for the purpose prove, or can be counted upon to prove, promotive and contributive forces towards development conceived and implemented on the nation-wide scale – or, alternatively, to which degree and in which manner they prove or can be reckoned as, stumbling blocks or "weak spots" to same. No doubt, an important consideration in this connection is the on-the-spot implementation – locally and/or regionally – of measures devised and taken at the "central" level: that is, in the more or less metropolitan centre of the country. For purposes of investigation an a priory typology of modalities in this regard is likely to prove most useful. If more than one research team would concern themselves with the matter such a typology could profitably be developed in mutual consultation for common use.

Research projects that would concentrate upon this viewpoint may either deal with an entire national context, or with a particular part of a country consisting of a number of (adjacent) regional and/or local units. They may further concentrate on the economic, or the social, or the political, or the administrative, or on a combination of two or more of these aspects. In all cases they will have to account for the manner in which development planning and development measures originating at the nationwide, "central" level conceive of the manner in which local and regional units add up to the nationwide unit: conception that tends to be a tacit presupposition in many cases and that does not show in its real fulness and complexity in constitutional documents and the like. Just as it was necessary, above, to underscore the importance of existing images of development, so it is important here to point to the existence of (a variety of) images of the operational pattern of national cohesion. Once more, a typology of images (it would rather closely parallel the typology just suggested without, however, being completely identical therewith) is likely to prove helpful for investigations.

b. Second, the viewpoint of the *local or regional entity as part of the nationwide context*; again: social, economic, administrative, political, also cultural. The view now moves from the component local/regional part towards the encompassing national framework. What appears as crucial here is thus primarily the local/regional conception as to the degree and mode of its integration into the wider context. This conception is determined by such factors as cultural, historical and geographical specificity, corresponding self-view and self-identification, whether as distinct from or yet as one component of the wider national entity. It is furthermore determined by the local/regional specificity in the prevailing definition of commonwealth and common interest, also in an often traditionally determined view concerning the relationships between power and authority in the local/regional context and same in the wider national context, or, specific variant of same, at a local/regional level and at a national level.

Research contributions meant to feed in at this point of the proposed project will usually be concentrated upon one regional or a few local units, and set out from an attempt to determine their specificity in regard to development needs and possibilities. Once these have been asserted, both per se and in the perspective of national development needs and possibilities of the nation-state of which the regional or local unit concerned happens to be part, the analysis can be pursued along some of the lines just suggested. In so doing, one may choose to concentrate upon certain aspects, but it may be hard to proceed exclusively in terms of one discipline. A rather more multidisciplinary or even interdisciplinary procedure would seen more promising.

c. A third angle of approach is provided by looking at the matter in terms of the *interaction between the two distinct and complementary viewpoints* just discussed. This is the approach that will discern the actual implementation of given policies and measures, and that in doing so will try to establish typologies likely to prove of heuristic value. No doubt, before one may hope to achieve a sensible typology, a number of limited case studies of this nature will have to be achieved. Most of these case studies will have to find their primary limitation in restriction to a particular local or regional scope, in which nationwide or perhaps specifically local/regional policies are pursued. Its secondary limitation will normally ensue from concentration upon particular measures and policies, or same in a particular field: be this economic, social, political, or be it a complex of measures like industrialization or agricultural reorganization. Naturally, research work of this nature needs to be descriptive to a considerable extent, but it is hoped that, unlike what happens only too often, it will not prove abortive by never moving beyond description. What we really need is analysis following upon description: analysis of the manner in which what is done in the nationwide context does or does not affect the local/regional entity, and of the manner in which that which goes on in the local/regional context is or is not relevant to the wider national framework, and in which way. It is only on the basis of such further analysis that one may hope to establish the operational typology just suggested.

SOME READINGS

Ansari, M. A. S. (ed.), *Social Research in National Development* (Peshawar, Pakistan Academy for Rural Development, 1963).

Beattie, J., *Other Cultures: Aims, Methods and Achievements in Social Anthropology* (London, Cohen and West, 1964).

Blaug, M., *Economic Theory in Retrospect* (London, Heinemann, 1964).

Cuisenier, J. (ed.), *Problèmes du développement économique dans les pays méditerranéens* (Paris/La Haye, Mouton, 1963).

Friedmann, J. (ed.), "Regional Development and Planning". Special issue of *Journal of the American Institute of Planners*, no. 2 (Berkeley, 1964), pp. 82-176.

Glikson, A., *Regional Planning and Development* (Leiden, Sijthoff, 1955).

Granger, G. G., *Méthodologie économique* (Paris, P.U.F., 1955).

Gusdorf, G., *Introduction aux sciences humaines* (Paris, Belles Lettres, 1960).

Heck, B. C. van, *Afbakening en typologie van homogene zones* (Assen, van Gorcum, 1965).

Isard, W. (ed.), *Regional Economic Planning: Techniques for Analysis* (Paris, EPA/OEEC, 1961).

Kuenstler, P., et al., *World, Nations and Groups in Development* (The Hague, Mouton, 1963).

Lévi-Strauss, Cl., *Anthropologie structurale* (Paris, Plon, 1958).
Mair, L. P., *Studies in Applied Anthropology* (University of London, The Athlone Press, 1957).
Martindale, D., *The Nature and Type of Sociological Theory* (London, Routledge Kegan Paul, 1961).
Nieuwenhuijze, C. A. O. van (ed.), *Intelligible Fields in the Social Sciences* (With chapters on "Middle Range Theory" and "Area Studies") (The Hague, Mouton, 1966).
Pitt-Rivers, J. (ed.), *Mediterranean Countrymen* (Paris/La Haye, Mouton, 1963).
Rossi Doria, M., et al., *Aspetti e problemi sociali dello sviluppo economico in Italia* (Bari, Laterza, 1959).
Ruopp, Ph. (ed.), *Approaches to Community Development* (The Hague, Van Hoeven, 1953).
SVIMEZ, *Sviluppo industriale e imprenditori locali* (Roma, Giuffrè, 1962).
United Nations, *Decentralization for National and Local Development.* (ST/TAO/M /19) (1962).
—, *Community Development and National Development.* Report by an ad hoc group of experts appointed by the Secretary General of the United Nations (New York, 1963).
Ward, B., *Problems of Greek Regional Development* (Athens, Center of Economic Research, 1963).
Zetterberg, H. L., *On Theory and Verification in Sociology* (Totowa N.J., Bedminster, 1963).

III

THE IDEAL CITY OR THE VARIETIES OF METASOCIAL EXPERIENCE

A Typology

> Unless the ideal build the house – and with it the city also – they labour in vain that build it.
>
> Thus, in all these ways, we will transmit this City, not only not less, but greater, better and more beautiful than it was transmitted to us.
>
> (Two quotations in Patrick Geddes, 1949, 100, viii)

1. What constitutes an orchestra is the variety of the instruments of which it is composed. Amongst the factors making for variety between instruments is the difference in richness of tone. There is no instrument that, when played, will produce sound of one wavelength only. Even when played so as to produce one particular tone, it will produce, besides, a quantity of sound of wavelengths different from, although often somehow related to, the intended one. But some instruments are noted for their pure, and, in a sense, meager tone: most of their sound output having the one intended wavelength. Others, contrariwise, produce a rich but far less pure or true tone. For example, think of the saxophone and some modern electronic devices on the one hand and of the kettle drums on the other. Clearly, one way of providing a characteristic of a given instrument is by analyzing how much of its sound output occurs on wavelengths other than the intended one and how these relate to the latter.

1.1. If it is permissible to compare the social sciences with an orchestra (some will certainly say that to do so is presumptuous on the part of the social sciences), then the various concepts used – whether technical or not so technical, whether theoretical or rather common-sense – may perhaps be compared to instruments. Some concepts have a thin and pure "tone". Others, not necessarily despicable on this account, have a rich but less pure "tone". In fact, amongst the latter one is likely to find some that produce an overwhelmingly broad range of tones in addition to the one for which one happens to play them. Amongst these additional tones, some will have a wavelength harmonically related to the intended one, others not. Certainly, the use of any given concept will be determined, one way or another, by its characteristics as these appear in the (metaphorical) terms just given. And part of any social sciences training must accordingly be, to learn to assess the use of a given concept in the light of its characteristics of tone: whether meager and pure or rich and overloaded. But this is not the point to be argued here.

1.2. One concept of the social sciences will be singled out for the purpose of this paper. Its tone is rich as that of a whole battery of kettle drums. In surveying and discussing it, an attempt will be made to assess its main and additional wavelengths, so as to gain a full perspective of its wealth in meaning and thereby of its use for theoretical and practical purposes. The concept is – the ideal city.

2. Before one may venture upon a discussion of the ideal city one must deal with the city. The concept "city" appears, to the analytic mind, to connote a reality at once ideal and empirical. Primarily, the city marks itself off, not – as we are inclined to think nowadays – from the countryside, but from barbarism, from "le vide culturel". Note that the word *civis,* from which derives *civitas,* the root of our word city, has an adjective *civilis,* from which is derived our word civilization.

Gregariousness and civilization are equally indicated in the one basic notion. The empirical city exists as an approximative implementation of an image that the citizens share. Conversely the ideal city is but the persistent transcendence of a spatio-temporal concentration of human beings. Appearing to one viewpoint as comparable to an organism in the biological sense, the city is, to a complementary viewpoint, experienced as organization. Thus it acquires, to the experiencing and reflecting mind, significance – significance that renders the human city richer, by one dimension of reality, than the anthill or the beehive. To the analytic mind the concept shows therefore dual aspects. Does this turn the concept itself into a dual, complex entity? Not in the sense that the complementarity of aspects could in any way affect the underlying consistent unity. In showing dual aspects the city but reflects the psycho-biological complexity of its component element – man. Human togetherness, once self-sustained in a concentration marking it off from encompassing nature, is a matter of significance in addition to being a matter of biological fact. Of these two aspects, the one always presupposes the other. As compared to anthill and beehive, the city is human in that these two aspects – its empirical occurrence and its significance – stand in persistent interaction. The ideal city *per se* is exactly as impossible, to human existence, as the empirical city *per se*.

2.1. Thus, a discussion of the ideal city simultaneously presupposes and anticipates a discussion of the empirical city, of the crude facts of human conglomeration.

2.2. It is now feasible to indicate in which meaning the term city will be used in the present paper. It is a meaning derived in etymologically valid manner from the Latin root of the word. It encompasses, on the one hand, the urban concentrations of people which are normally denoted by the term cities, in current usage. On the other hand, it is also meant to include any organized society built around one or more towns with surrounding countryside, in which one sociocultural entity achieves its own self-realization: a meaning still found in the French word *cité*. (For example, the title of L. Gardet's work, *La cité musulmane*, 1954, which would have to be translated as Islamic society. It recalls that famous classic, Fustel de Coulange's *La cité antique*.) Given the stress on the ideal rather than on the empirical aspect of the city that will prevail throughout this paper, it will prove possible and even inevitable to use the two shades of meaning almost indiscriminately. For most of our present purposes, they are virtually synonymous.

3. Rogert Mucchielli has entitled his recent study, *Le mythe de la cité idéale* (1960). The title provides a valuable clue for the study of the ideal city: in its interrelationship to the empirical city. Myth is that which signals a reality that transcends phenomenal existence. It renders this transcendent reality intelligible by means of names and events that are withdrawn from the spatio-temporal here-and-now. Thus what is told as myth is not a factual account of phenomenal reality. But to the extent to which it deviates from being a factual account, it renders the more important service of being an elucidation, an attribution of sense to phenomenal reality. In myth, man holds up a mirror to the phenomenal reality in which he exists – and more than a mirror: a model, an image. The truth of the image is not in the mechanical accuracy with which it reflects: much rather it is in the degree to which it conveys sense, purpose, significance; the degree to which it humanizes – should one dare say, spiritualizes – phenomenal existence.

3.1. The relationship between myth and phenomenal reality, if inspected for its own sake, thus shows a complementarity of aspects.

In myth, reality stands affirmed and, what is more, confirmed. But in myth reality stands judged; and that may mean, rejected. The two indictments, positive and negative, are simultaneous and do not cancel one another out. But the one may, in a particular case, receive more stress than the other, or become more effective in regard to practical consequences. According to which of the two positions proves how much more important than the other, a range of situations may be distinguished. This is thus a range of varying correspondences between ideal city and phenomenal, empirical city, between myth and phenomenal reality. Were it not that the term is too categorical, too simplistic, one could speak of the degree of congruence between the two. So much is clear that on one extreme of this range stands self-defeating contradiction, and on the other extreme self-sustained harmony. It does not pay, at this point, to offer an enumeration of specific examples, inasmuch as a valid discussion of examples cannot take place until after the introduction of a second determining range. One notes, however, that, given the nature of either extreme on this scale, the more normal positions are likely to occur around the half-way mark, in between the extremes. These are positions where between the city and its image, between phenomenal and ideal society, a noticeable yet bearable degree of tension exists. Motoric tension, that in itself is necessarily constructive and disruptive at once: again with the stress more one way or more the other way. This, however, refers already to the second range, to be introduced below.

3.2. The persistent interaction between empirical city and ideal city has thus one preponderant modality. One can present it as a range of degrees of "congruence". To simplify the use of this range for the purposes of this paper, it may be conceived as running between two extreme positions, of which one could be called confirmation, the other rejection. The confrontation between ideal and phenomenal city may either entail the confirmation of the latter or its rejection; but usually the judgment will end up somewhere in the balance between the two extremes. (See Fig. 4.)

4. The congruence-contrast range must be distinguished from a second range of possible positions. This second range encompasses the various attitudes, on the part of those humans effectively composing the city concerned, that may reflect their awareness as to the particular position in the congruence/contrast range that happens to prevail amongst them. Their attitude and behaviour in regard to their city stands to be in-

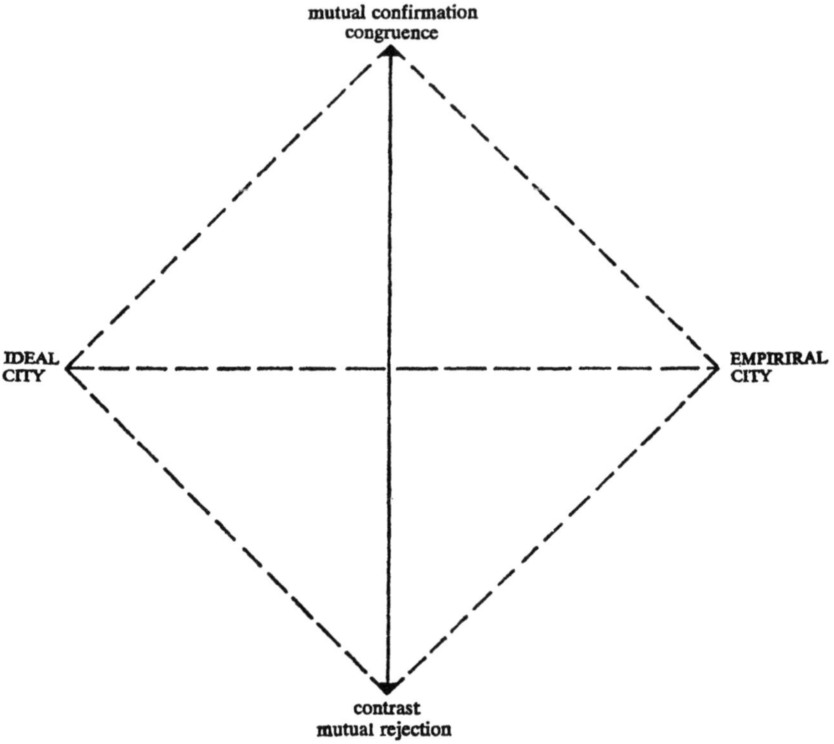

Fig. 4. Congruence/contrast range: the range of positions ensuing from confrontation between ideal city and empirical city, from "congruence" to contrast.

fluenced, if not determined, by their awareness concerning the relationship between ideal and phenomenal city under which they happen to live. This second range represents the impact of the former range, according to instances in which it occurs.

4.1. This attitudinal range, in its turn, must first be indicated by means of its two extremes. A first approximation to the nature of these extremes may be found in the terms constructive and destructive. Under the impact of what they find to be the relationship between their ideal city and their phenomenal city, people may tend towards either a constructive or a destructive attitude towards the prevailing amalgam of ideal and phenomenal reality: more particularly towards the actual, mostly phenomenal conditions under which they live. By way of further approximation, one may indicate the range of responses to any positions in the congruence/contrast range as runnning from positively functional on one extreme to dysfunctional on the other. A third ap-

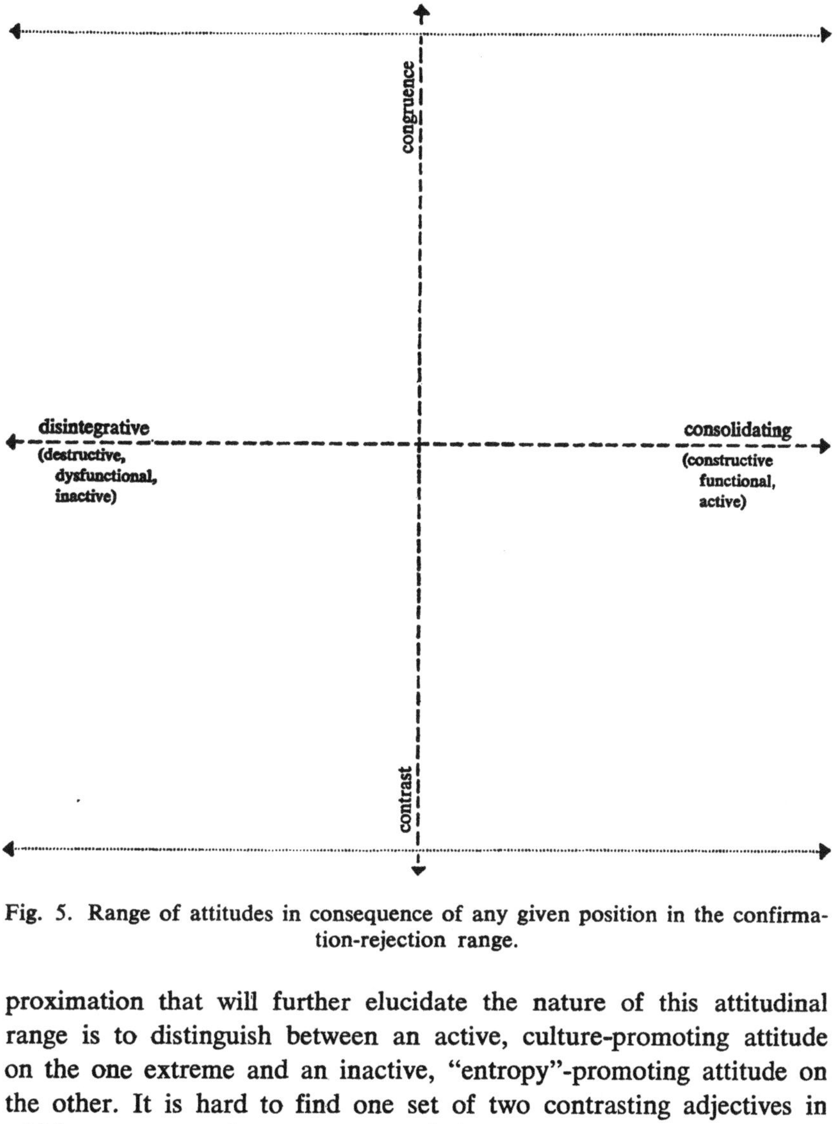

Fig. 5. Range of attitudes in consequence of any given position in the confirmation-rejection range.

proximation that will further elucidate the nature of this attitudinal range is to distinguish between an active, culture-promoting attitude on the one extreme and an inactive, "entropy"-promoting attitude on the other. It is hard to find one set of two contrasting adjectives in which to sum up the true nature of this range. After what has been said, it may be sufficient for purposes of short reference in this paper to call the one extreme on this range consolidating, the other disintegrative.

5. For purposes of further elucidation of the two ranges, combined as in Fig. 5, four additional remarks are necessary.

5.1. The two ranges together, as sketched in Fig. 5, provide a continuum, fit for location of positions and thus capable of serving as the frame of reference for a typology of ideal city phenomena. It is possible, with this frame of reference, to construct a diagram showing differences and interrelationships between the numerous and considerably varied phenomena that come under the one heading, ideal city. (See Fig. 11, p. 141.)

5.2. The two ranges have in common that they are not simply quantitative ranges, each indicating one category in a scale from infinite negative to infinite positive. On the contrary, each range embraces a complementarity relationship of two categories, each of which occurs from one extreme of the range to the other. What makes for the different positions in the range is the "mix" of the two complementary elements, the degree of effectiveness with which the one obliterates the other. Thus, although the ranges are constructed in a way deceptively similar to simple quantitative ranges each showing one qualitative category only, they actually show a qualitative category that consists of, and varies with, the "mix" of two complementary qualitative categories. The "mix", in its turn, is not a matter of simply more or less, but a matter of preponderance as between the two complementary categories concerned.

5.3. One could wonder whether the continuum thus sketched holds provision for what seems to be an important turn of events and one of reasonably frequent occurrence, viz. interruption in the interaction between a given ideal city and its corresponding empirical city. The rate of change in the one may be so different from that in the other that the correspondence is affected in a measure that cannot be coped with within the limits of the existing confirmation-rejection range. Or a number of those composing a given empirical city may shift allegiance from the hitherto generally prevailing ideal city concepts to a competing set of ideal city concepts. The latter eventuality would appear even more detrimental to the use of the proposed frame of reference than the former. It need not be, however: if one is willing to admit that the frame of reference can be used in repetitive fashion, not merely from case to case of given combinations of ideal and "corresponding" empirical city, but equally well in cases of more than one ideal city to refer, perhaps on a competitive basis, to one empirical city, – or what happens as well, the reverse position of more than one empirical city supposedly corresponding to one and the same ideal city.

5.4. The various positions that can be sketched in the proposed

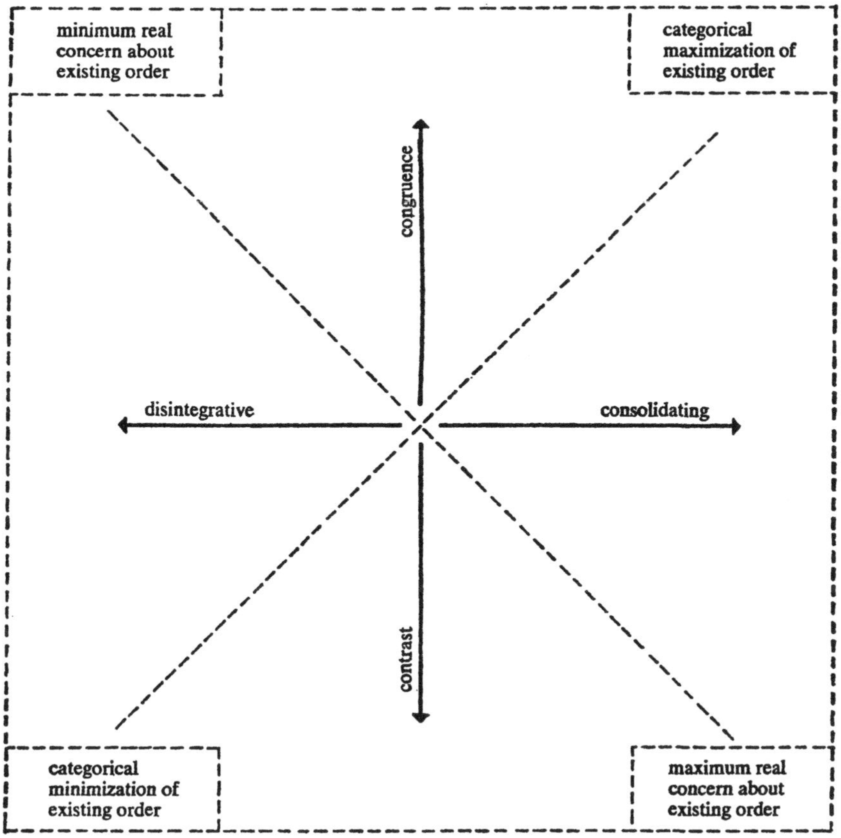

Fig. 6. Main trends according to which types of ideal city conceptions can be distinguished.

frame of reference must obviously find their place in accordance with the composite result of placement according to either range. It is interesting to note that opposite positions in this framework do not merely show the impact of complementarity when considered across either of the two ranges, but also when considered diagonally. Consequently, it is possible to distinguish four main trends according to which positions can be ordered in the proposed typology: each of them standing in one corner of the diagram presented in Fig. 5. (See Fig. 6.)

5.5. It is now proposed to arrange the wealth of material that demands attention, in such a manner that the possibilities offered by the typological scheme are concentrated into four main headings. These headings are found by imposing, upon the given analytic frame of reference, four different qualitative denominators. Each of these will then naturally show a number of distinct positions or variants, each in its turn with sub-positions or sub-variants.

By providing cases in point to demonstrate a reasonable number of these, it will be possible to show, at once, the systematizing possibilities of the scheme now developed and the ramifications, in form and content, of that complex thing which is known by the amazingly simple name of ideal city.

5.5.1. The headings to be employed are the following. First (Fig. 7; positions I^A, I^B, I^C, I^D and one position left blank) – pragmatist approaches to which the ideal city is mainly the manifestation of an intelligibility inherent in the phenomenal order of things. Prevailing state of affairs and prevailing philosophy coincide, whether virtually or actually, so as to bring ideal city and phenomenal city very close together. (See p. 83.)

Secondly (Fig. 8; positions II^A, II^B) – normative approaches, determined by a priori established religious or philosophical frames of reference. The ideal city features as the norm for the phenomenal city and as such is fundamentally inherent in it. (See p. 93.)

Thirdly (Fig. 9; positions III^A, III^B, III^C, III^D) – mythical-idealist approaches according to which the ideal city is somehow of a different, higher (whether moral, logical or otherwise) order than the phenomenal city. Its relative transcendence entails strikingly varied consequences; as a result, the graphic presentation of this variant assumes a somewhat scattered appearance that is hardly fit to represent its basic consistency of type. (See p. 111.)

Fourthly (Fig. 10; position IV and counterpart position) – the attitude that is out for drawing practical consequences from whatever it perceives in the way of differences between ideal and phenomenal city. Compared to this fourth attitude, the three other ones are speculative; on the other hand, they tend to be clearer and more articulate in regard to the ideal city: *per se* and in its relationship to the phenomenal city. (See p. 128.) For a combined drawing showing all four headings see Fig. 11, p. 141.)

6. The first kind of positions to be distinguished in this typology (Fig. 7) is distinct by the manner in which the ideal city concepts feature as rationalizations, as manifestations of intelligibility inherent in the phenomenal order. The ideal city is, then, the point of coincidence between the prevailing state of affairs and the prevailing philosophical pattern of a given area and period. The visual representation of this

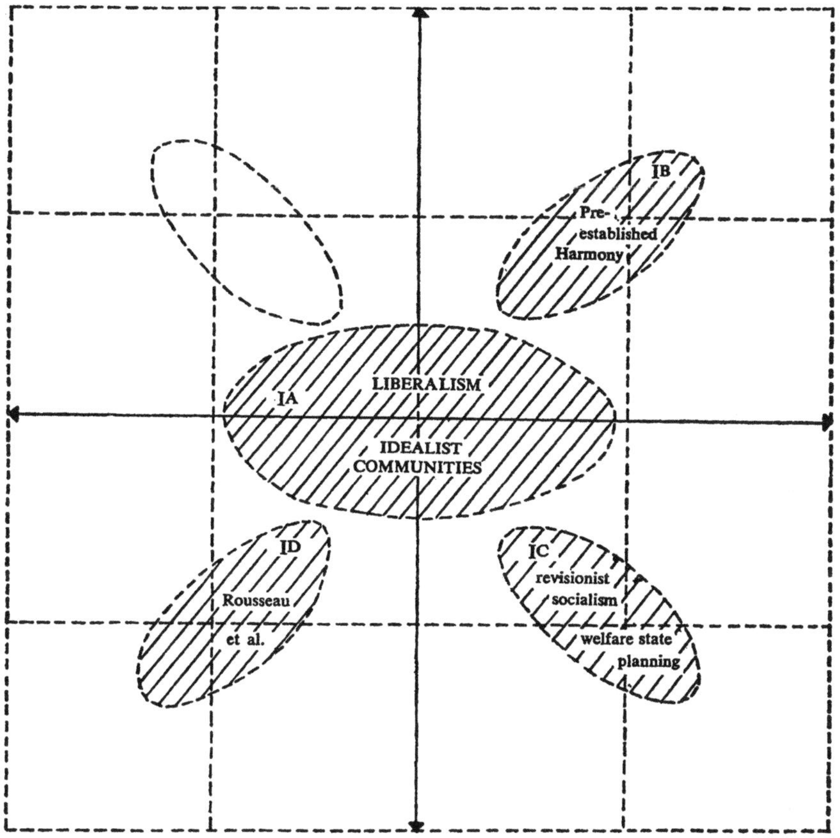

Fig. 7. Predominantly pragmatist approaches, not without philosophical underpinning: trends of socio-economic action.

type is somewhat curious, but one understands it upon closer consideration. One variant (position I^A, Fig. 7) is uncommitted in regard to any of the determinants employed in establishing the present typological scheme. It features thus in the central, zero area of the drawing. But then, there are four further variants, two of them more and two of them less committed to the prevailing state of affairs. One of the former two (position I^B, Fig. 7) is committed in a conservative fashion; another (position I^C, Fig. 7) in more revisionist fashion. Their graphical presentation assumes the shape of two offshoots to the first variant, one leaning to the right top corner, the other to the right bottom corner.

One can add offshoots on the left side so as to establish a symmetrical presentation. Thus, one may build out toward the left bottom corner position I^D, Fig. 7. In this position one may bring together types of thought resembling on the one hand what has here been distinguished, under position III^B, as revolutionary thought and on the other hand what has here been styled, under position II^B, utopian thought. The affinity between these two types, insofar as it could find its reflection in the present context (constructs of virtual identity between ideal and phenomenal order), would then perhaps come out more clearly as a matter in its own right. Mucchielli (1960) applies a similar procedure. For present purposes, it is proposed not to go into too much detail in elaborating this particular position; less so since Mucchielli's material on the matter is readily available. For the sake of simplicity, one case in point, namely Rousseau, may suffice.

In the same fashion again, one could consider dealing in a position towards the left top corner with a type that will here be treated as one sub-variant of position III^C (Fig. 9): the culture pessimists of whom Spengler (1927) is the most famous though perhaps in retrospect not the most convincing example and with whom Nisbet (1962, 7) also counts Toynbee because of what he suspects to be his preoccupation with patterns of decay.[1] However, in order not to overload the scheme with substantiating material, it is proposed to refrain from filling in this fourth corner position that should render Fig. 4 fully symmetrical.

6.1. Position I^A (Fig. 7) can be exemplified in liberalism of the *laissez-faire* type.[2] The liberalism, more precisely, that upon having achieved liberation from traditional power envisages phenomenal reality in the rosy light of an optimism that will ensue from the conviction that the present state of affairs is exactly as it should be. Phenomenal order and natural order, by coinciding, are hallowed as ideal order.

"Early liberalism was optimistic, energetic and philosophic, because it represented growing forces which appeared likely to become victorious without great difficulty, and to bring by their victory great benefits to mankind". (Russel 1946, 64).

Speaking of the spirit of *laissez-faire*, Schumpeter (1954, 394) adds, "that is to say, (. . .) the theory that the best way of promoting eco-

[1] A similar difficulty in regard to placement occurs in position I C as compared with position III D. Planning for welfare and development could be ranged with either position inasmuch as in most cases one has no way of distinguishing the mythical-idealist content of a planning conception from its "intelligible model" content. A very precise distinction should range the former with position III D and the latter with position I C.

[2] The saying *laissez faire, laissez passer* is attributed to De Gournay (Schumpeter, 1954, 244 note 1).

nomic development and general welfare is to remove fetters from the private-enterprise economy and to leave it alone". Once free in its movements, the entrepreneurial order thus achieves the full coincidence of phenomenal and ideal order.

Adam Smith (1776, Vol. I) marks the point of coincidence in his own manner. In his basic work on economics, he sets out from the division of labour (mainly with regard to industry) as a crucial notion (p. 7), complemented with the saving of labour (p. 16). He ascribes it to "a certain propensity in the human nature" (p. 20) which is furthermore ascribed, in people, "not to their humanity but to their self-love" (p. 22). Innocent propositions which actually imply no less than a virtual identification between the essential, necessary data of existence and the actually prevailing economic order of the day, which, as happens so often, is seen as the economic order writ large. Surely, in the margin that remains between virtual and full identification, there exists ample latitude for theorizing even about several different economic systems. But the categorical imperative (without the name) is established and all one has to do, basically, is to allow the natural order to assert itself.[3]

6.1.1. There is yet another kind of phenomenon, apparently very different from liberal doctrine, that deserves to be mentioned as a case in point to demonstrate the present position. The ideal and the real may coincide in such a manner that the ideal city is just simply there: *vorgefunden*. They may also coincide because one makes them do so. This happens when a group of people do what Plato achieves in imagination only – leave one phenomenal order that is not to their liking an, having left it to the extent that to them it exists no more, establish their own preconceived ideal city in what is otherwise empty space: starting from a clean slate and without outside interference. Mucchielli (1960, 177ff.), who for unclear reasons identifies these with the reformers, lists an impressive number of the well-known communities of idealists and discusses them from the viewpoint of their psychological, ethical, economic and constitutional characteristics (p. 119). This scheme seems somewhat to soft-pedal the religious dimensions, in particular the salvation element; particularly so in cases when the ideal community is established on the strength of (sectarian) religious ideas, such as the Shakers (p. 127ff.).

[3] Idea that, in different elaboration, is traceable even in Marxist revolutionary theory, and also in the preoccupation of revisionist socialism to eliminate obstacles on the way to betterment that have been inherited from the past.

In his summing-up (p. 145) Mucchielli has limited himself to the 19th century cases.[4] It is perhaps enlightening to add two contemporary offshoots, representing exactly the same position, the one in most ephemeral fashion, the other with rigorous consistency. The former is a suspect if not outright disreputable heir to the idea of the ideal city here-and-now away from all cities: the nudist camp. Ideal in a naturistic, quasi-paradisiac fashion, it does hardly lend itself to quotation in terms of ideas.

The other is perhaps the most discussed and glorified community in the entire world: Israel, the twice promised land and to the ingathered part of the Jewry as much as to most of those who remain in diaspora the actually existing ideal city; with its small-size, community-level replica, the *kibbutz*. The latter, moreover, is also a direct spiritual and material descendant of the romantic communities discussed by Mucchielli. After the facts, the former has even been given its myth, or at least its mythical-epical literary projection, in Leon Uris' bestseller *Exodus*. Israel is not nearly so conspicuous for what it is becoming or hopes to be: it is considered for what it *is*, what it represents.

Joseph Baratz, in telling why he opted for a course that would eventually make him and his fellows the creators of the *kibbutz*, speaks of the means by which "the Jews were to be regenerated. What we wanted was to work ourselves, to be as self-supporting as we could and to do it not for wages but for the satisfaction of helping one another and of tending the soil. We knew that we needed one another's close and constant help because in the harsh conditions of the new country neither a person or a family could stand alone, and it seemed likely that if we worked more closely together we should produce more. But we were not only thinking of producing big crops, though this was important since the country must support more people; the fact of producing – growing things – had for us a meaning which was related to the whole of human life. It was this wholeness we had lacked in exile. There we had been cut off from nature, from our roots, and everything had been distorted by the need for security – even the family had become a little fortress. We had had to buy security with money and make money with whatever are the money-making faculties of the brain, we had lived on our brains and we hadn't used them

[4] He offers brief, lucid profiles of the Guarani republic of Paraguay (1588), of the Shakers (1770), of the Phalanstère of Fourier (1832). (Years indicate beginning of movements.) Compare also Maunier (1949), Ch. 16.

properly because we had been cut off from the growing, natural side of life. That was why we had become so dry, so barren" (1954, 44f.).

"Such then, is the common ground of the generations of men called Jews. Whatever part of the world they live or die in, however diverse their looks, their speech, their ways and works, they are made one by the belief, passed from father to son, that Palestina and no other land is the land of Israel, that to live elsewhere is to live in exile and dispersion, that their destiny is to regather and return" (Kallen, 1958, 25). (For a less intuitive narrative which however shows the present point of interest only by implication, see Halpern 1961. For some badly needed and brilliantly executed soul-searching on the fundamental issues of nationhood involved in all this, see Friedmann, 1965.)

6.2. The identity that, to the viewpoint now under discussion, is taken to obtain between ideal and phenomenal order, once conceived in a historical perspective, will tend to appear either as virtual identity, *i.e.*, as identity within reach yet to be implemented, or as consummated identity, *i.e.*, as identity to be maintained. Given the fundamentally non-historic nature of the construct, the two aspects remain undistinguished and almost undistinguishable in many cases. Here and there, however, one encounters cases in which the one has decidedly more accent than the other. These are the cases graphically presented, in Fig. 7, by the offshoots from the main oval that point in the four diagonal directions of the scheme. (Obviously four directions, since the dual appearances of identity just mentioned may either be deemed desirable, or appear as repugnant.) As stated above, it is not the intention here to discuss all four of them; but some illustrations are certainly called for.

6.3. Position I^B (Fig. 7) may be exemplified with one crucial idea of one philosopher: Leibnitz's pre-established harmony. Russell (1946, 604ff.) distinguishes sharply between two Leibnitzes, "the popular Leibnitz who invented the doctrine that this is the best of all possible worlds" and another one who "left the records of his thought unpublished in his desk". But in either Leibnitz, the one who upholds the prevailing order so as to gain favour and the thinker independent of the prevailing order, the concern is with the essence of reality.

He conceives it as an aggregate (Leibnitz, 1964, 69 comp. 27) of substances. The substances are so-called monads, a kind of physical-metaphysical atoms (*ibid.*, 69), which he considers as souls (*ibid.*, 81): immaterially. Monads do not interact, they are "windowless" (*ibid.*, 71); yet together they are subject to (hierarchical) order and change (*ibid.,* 73f.). God, agent (more precisely, sufficient cause: *ibid.,* p. 93) of the aggregation (prime unity: *ibid.,* p. 97, comp. 107f.) which turns the infinite number of monads into comprehensive reality, imposes (or should one say, appears as) pre-established harmony (*ibid.,* 119). The popular Leibnitz then elaborates this doctrine so as to show that the ensuing reality is the best of possible worlds (Russell, 1946, 612f.). "The conclusion is easy that the aggregate of all souls must constitute the city of God, *i.e.,* the most perfect state which is possible under the most perfect monarch. This city of God, this truly universal monarchy is a moral world in the natural world (. . .)" (Leibnitz, 1954, 123). The esoteric Leibnitz, employing a similarly deistic, residual notion of God by way of ultimate philosophic category, ends up in rigid determinism (Russell, 1946, 616).

Thus the phenomenal order is necessarily identical with the ideal order: both in the sense that the ideal city renders the phenomenal order intelligible, shows its real being, and in the sense that the real order is the necessary counterpart to, manifestation of, the ideal order; – in terms that will not endanger the philosopher's position, the best of possible worlds.

6.4. Position I^C^ (Fig. 7) is represented by what Mucchielli (1960, Ch. IV, p. 104ff.) calls reforms: the attitude of the reformist which "always supposes a 'plan' and the plan in its turn supposes a notion – prior to a forecast or estimate – of the means. Every reformer has a political system that he seeks to realize to a greater or lesser extent, in combat with the successive obstacles he will encounter" (p. 104). He distinguishes "two kinds of reformers; on the one hand those who, armed with a political system, undertake or propose the reform action and on the other hand the appeal to those in power made by utopians, in order to make them adopt the grand scheme" (p. 106).

Indeed the affinity to the utopian position (II^B^, Fig. 8) is clear; so much so that for present purposes it is more convenient to concentrate upon the former of the two, relegating the latter to the utopian position.

"The reformer is interested in evolution and thus in the world as it is. This is why he seems always more 'economical', more 'realistic' and more 'evolutionist' than the author of a utopia" (p. 106). In fact, the difference may well lie deeper. To the reformer, phenomenal reality and ideal city are much less than to the utopian of different orders. Between the two, the reformer envisages a distinct, not just imaginary link of (guided) evolution taking a foreseeable span of time; the utopian sees little but vain hope and despair.

Besides the older examples of reformers [5] quoted by Mucchielli (p. 107ff.) there are more modern ideas that can serve equally well, if not better, as cases in point to demonstrate the present position. One of these is the famous Beveridge report (1942). In its abbreviated version it lists three guiding principles (p. 4f.).

"7. The first principle is that any proposals for the future while they should use to the full the experience gathered in the past, should not be restricted by consideration of sectional interests established in the obtaining of that experience. Now, when the war is abolishing landmarks of every kind, is the opportunity for using experience in a clear field. A revolutionary moment in the world's history is a time for revolutions, not for patching.[6]

"8. The second principle is that organisation of social insurance should be treated as one part only of a comprehensive policy of social progress. Social insurance fully developed may provide income security; it is an attack upon Want. But Want is one only of five giants on the road of reconstruction and in some ways the easiest to attack. The others are Disease, Ignorance, Squalor and Idleness."

"9. The third principle is that Social Security must be achieved by co-operation between the State and the individual. The state should offer security for service and contribution. The state in organizing security should not stifle incentive, opportunity, responsibility; in establishing a national minimum, it should leave room and encouragement for voluntary action by each individual to provide more than that minimum for himself and his family."

The same tendency to envisage the ideal city as a step-by-step improved version of the phenomenal city lies at the root of revisionist – as distinct from, in fact, opposed to revolutionary – socialism. The ideas that determine the stand taken by labour parties are characteristic

[5] Mucchielli mentions reformers such as Peter Chamberlen (1649), Gerrard Winstanley (1652), Peter Cornelius Plockhoy (1659), Robert Wallace (1761), James Harrington (1656) and John Bellers (1695). (Years refer to publication of book serving as illustration.) A listing of more contemporary phenomena should include such movements as the Fabian Society along with individual thinkers.

[6] A somewhat unfortunate sentence, given the illustrative purposes for which the passage is quoted here. By the words of this sentence, one could be induced to wonder whether some illustration to the revolutionary attitude has perhaps stranded here, instead of being used at its proper place. Such is not the case. The revolutionary noises made are mere rhetoric used somewhat out of proportion to meanings conveyed.

in this regard. The Dutch Labour Party, for one, has published an elaborate document (significantly, though without justification, called a plan) outlining in detail how the future – read: ideal society – is to look as compared with the prevailing one and how to reach it in a gradual procedure. The title is, *The Road to Freedom: A socialist Perspective* (Plancommissie Partij v. d. Arbeid, 1951). The document is meant to provide a concrete elaboration of party aims (p. 5). Its opening statement, a signal statement for present purposes, is a quotation from the 1947 party platform, "In a world, shaken and dislocated by two wars, scourged by economic crises and ensuing mass unemployment, horrified by totalitarian terror, – which now, after the end of World War II, is deeply aware of disappointment caused by the failure of an ordered world community to materialize and by the threat of new war hazards and the burden of economic chaos, – millions are moved by the desire for an order of social justice, security of living and national well-being, spiritual freedom and world peace" (p. 7). The document then continues to say that against the background of so much frustrated hope the desire was to outline which way one would want to go for the coming years (*ibid.*).

"This plan is meant to constitute a *prise de conscience* of socialism in a critical period of history" (p. 8). Taking its point of issue from human freedom, in service to society and in responsibility (p. 9), the report finds many obstacles hampering the implementation of this goal, many of them inherited from the past (p. 11). A five-point plan is then developed, in regard to which it is furthermore said that "In the plan, whilst indicating the road to be pursued, reference is repeatedly made to planned guidance of social development" (p. 13). "Basic to the plan is the consideration that the main task of our period is to eliminate the obstacles that hamper man's development in society and to provide freedom with a firm foundation in responsibility through the creation of institutions (*levensvormen*) which hold people together in such a manner that to live in this society is experienced as both an individual and a common value. The means that are available towards the realization of this aim, are further elaborated (. . .)" (p. 15).

After the introductory statement, the book opens with a treatise on "social order in socialist perspective", which inevitably sets out from a discussion of property relationships.

6.5. Position I^D (Fig. 7), finally, will be exemplified, one for all, by J. J. Rouseau – more particularly by some passages from his *Du Contrat Social,* published 1762 in Amsterdam. As Guillemin points out in his introduction to the 1963 edition, "the Social Contract is not properly intelligible separate from the Discours, the Emile and the Héloise, but it is when considered in the framework of the development of all of them together. Then, we recognize Rousseau's leading ideas. The Contract is their application in regard to the set-up of the city" (p. 19). As to these general ideas, "the destination of creature is spelled out in book V of the Emile: serving God by doing what is good" (p. 11): never mind the Encyclopédistes. "Man, learn what God has ordained that you shall be, know your identity" (p. 12). He makes much of nature, and one of his biographers, Parodi, points out that "the idea of nature, with Rousseau, does not become fully clear without reference to religious ideas. Nature is, to him, the expression, in the form of primitive instincts, of God's will and intentions in our regard" (p. 14). Hence, "the form according to which men are governed is by no means indifferent in regard to those essentials which constitute their this-worldly and other-worldly destination" (p. 19). "People are what their government makes them to be" (p. 20).

Given such presuppositions nothing is more natural than to scrutinize the prevailing order down to its very roots. Thus the prevailing order and the natural order (but natural in a sense loaded with ideal city meaning) are confronted, with the latter serving as the yardstick by which to measure the former. And once the argument is under way, the former is present only by implication, as standing under the verdict of the latter, as virtually annihilated and replaced by the latter. "Established order? No, established disorder" (p. 29, annex note 36, p. 42).

"I propose to seek whether, in the civil order, some rule for legitimate and reliable administration can exist: taking people for what they are and laws for what they can be" (p. 49). "To find a form of association that defends and protects with all might the person and goods of every associate, and through which every one, in being united with all others, nevertheless obeys himself only and remains as free as before": this is the fundamental problem of which the *social contract* provides the solution" (p. 61). To be sure, social contract in a sense and with implications considerably more revolutionary than the way Locke uses the term. But, as Russell suggests, what turned the book into a revolutionary document was its condemnation followed by its adoption on the part of those who, in leading the French revolution, were apt to consider it as their Bible (1946, 727). (See, however, J. McDonald 1965.) Starting from the city or state thus established (p. 62), the natural consequence is the complete remaking, on paper, of society, seen from its political aspect. A remaking that turned out to demand the revolutionary attempt at remaking in reality.

7. The second range of variations on the given theme of attitudes in regard to the ideal city has as its common denominator the normative element. The ideal city features as a norm in regard to the phenomenal city; more precisely, a norm fundamentally inherent therein. The positions must therefore be drawn astride the vertical line that goes through the centre of the scheme.

One may furthermore distinguish between two kinds of positions within this range, according to whether the norm is assumed to be complied with (whether with or without a certain lag), or alternatively the norm is assumed to be upheld in contradistinction from phenomenal reality, thus to expose the insufficient measure of compliance on the latter's part. (The two positions have been sketched, in Fig. 8, as separate; one could also draw them in such a manner as to show a measure of interconnection. The difference is unimportant if considered in connection with the total picture that can be assembled from figures 7, 8, 9 and 10 together: see Fig. 11.)

Within either position one may distinguish various cases in point and assign to each of them places that are either nearer to the centre of the picture or nearer to the outside, whether top or bottom. The difference in placement, seen from top to bottom, represents the range from fully effective to fully questionable (and questioned) inherence of the normative ideal city in the phenomenal city. This section will again provide a random choice of cases in point. Insofar as the order of presentation could be taken as an order of arrangement in the scheme, it is proposed to work from top to bottom.

7.1. Position II^{A} (Fig. 8) represents the approach that takes the ideal city to be inherent in the phenomenal world and that assumes compliance on the part of the latter in regard to the former to prevail: to variable degrees.

7.1.1. The first case in point is the Chinese concept of *tao*. A useful discussion of this concept for present purposes is Fung Yu-lan's *The Spirit of Chinese Philosophy* (1947). The translator of this work has wisely added a summarized section from an earlier work by the same author, by way of introduction. The universe is there said to be distinguishable into four spheres of living (unselfconscious – utilitarian – moral – transcendant): an indirect but clear way to warn the Western

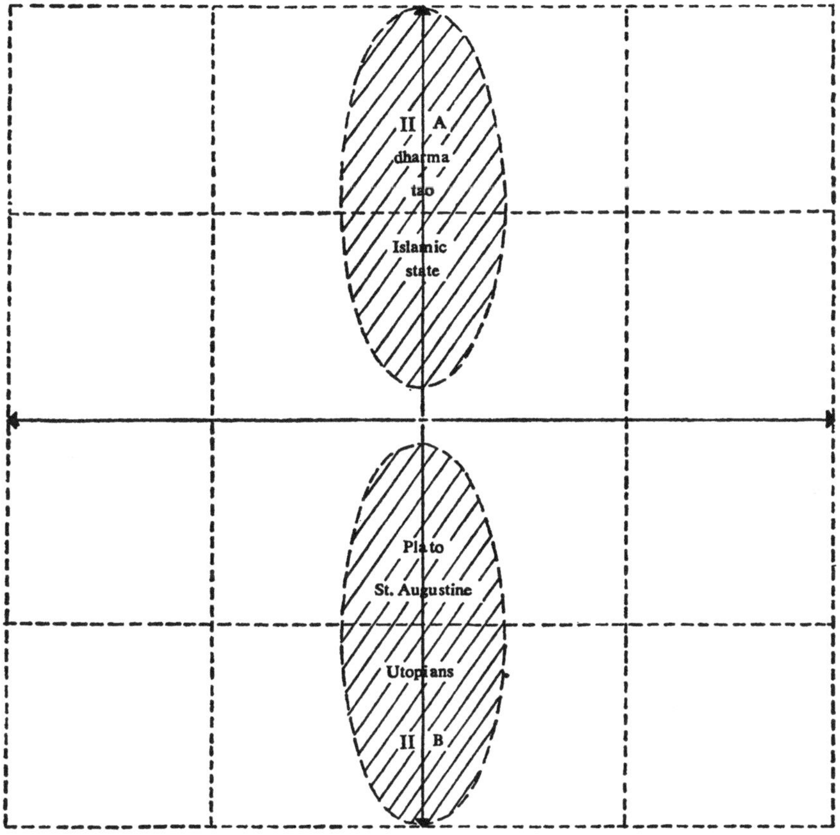

Fig. 8. Social philosophies linked up with more or less institutionalized religious or religio-philosophical systems – normative ways of life.

reader that the Western dichotomy between phenomenal and supernatural world is irrelevant to the universe of Chinese thought. For present purposes this means that the notion of the ideal city is irrelevant if it has to presuppose a basic dichotomy between an ideal and a phenomenal order of things. This however does not exclude the possibility to look for a Chinese counterpart to the ideal city complex, fitting in the frame of reference of classical Chinese thought.

In his introduction the same translator says about Chinese philosophy that, "so far as the main tenets of its tradition are concerned if we understand them, they cannot be said to be wholly other-worldly. We may use a newly coined expression and say that this philosophy is world-transcending. The meaning of this is that it is both of this world and of the other world. Chinese philosophy has one main tradition, one main stream of thought. This tradition is that it aims at a particular kind of highest life. But this kind of highest life, high though it is, is not divorced from the daily functioning of human relations. Thus it is both of this world and of the other world, and we maintain that it "both attains to sublime and yet performs the common tasks" (p. 2f.). And further, "The philosophers of China hold that the highest life of all, that at which philosophy aims, is both this-worldly and other-worldly; and that the men who are in possession of this highest life are the sages"; and, "Since the character of the sage is one of sageness within and kingliness without, philosophy, according to the Chinese tradition, is a branch of learning which exists to enable men to possess this kind of character. Therefore what philosophy discusses is what the philosophers of China describe as the Tao (Way) of "sageness within and kingliness without" (p. 4). Significantly, he quotes an older anonymous author in an attempt to suggest the scope and thereby the depth of Tao. This author "uses the expression 'sageness within and kingliness without', and denotes this principle as 'the Tao method', *i.e.* the method which enables men to know the whole truth. He holds that among the philosophers of his time not one obtained 'the Tao method' as a whole. Each only obtained part of it, or one aspect. Since each only partially apprehended the Tao method, therefore what each had to say was 'only the words of one school' and did not constitute an expression of the Tao: it was but one method with its limitations. Since the Tao is the principle of 'sageness within and kingliness without', it is also, from the point of view of its comprehensive use by man as a method, 'the attainment of the sublime and the performance of the common task' " (p. 5). Elsewhere in his argument, the author lists a number of terms as virtually synonymous, "we spoke of the universe, *i.e.,* 'the Great Whole', 'the essence', 'the world of the essence' and 'the Tao': concepts which are all philosophical concepts" (p. 23). This provides another view of the specifically Chinese context in which to distinguish the parallel to ideal city ideas.

By way of provisional profile to introduce his chapter on *tao*, which he considers as part of the general Chinese thought pattern, Granet (1934, 301) states that, "At the bottom of all conceptions of *tao* stand the notions of order, totality, responsibility, efficacy". He feels that in its early usage the word *tao* is employed "to express the effective order that dominates, as undetermined power, total phenomenal reality whilst itself remaining resistant against any determined realization" (p. 302).

But those thus using the concept, he adds, tend to illustrate it by referring to "the total art that allows a chief (. . .) to rule world and empire. Being the sole principle of success, *tao* becomes, to them, identical with the art of governing" (*ibid.*). The broader perspective thus evoked shows more clearly where he says, "The category of order or totality is the supreme category of Chinese thought" (p. 24). Also, "The notion of Tao is beyond the notions of force and of substance, and Yin and Yang, which feature indiscriminately as forces, substances and kinds, are something different again since these signs function in such a manner as to simultaneously classify and enliven the opposed aspects of the universal order: Tao, and Yin and Yang, evoke in synthetic fashion, engender in global fashion, the rhythmic order that presides over the world's life and the spirit's activity. Chinese thought seems entirely determined by the joint ideas of order, totality and rhythm" (p. 25).

The holist perception of the universe that appears to lie at the bottom shows in many ways. Thus the virtual identity that is conceived to prevail as between sign and signalled reality (p. 21f.); the refusal to "distinguish between the logical and the real" (p. 24). Thus the view according to which man and nature together are one *society* and which makes the Chinese conceive of an "etiquette of life" as a means to "implement total order" rather than, as Westerners would, of "a science aiming at knowledge of the world" (p. 24). Thus the Yin and Yang complex which, would Granet have had the contemporary concepts at his disposal, he might well have described as a complementarity construct denoting the operational aspect of the fundamental universal order (comp. Livre II, Ch. II, esp. p. 146f., 127): order which, summed up as one essential principle, bears the name *tao*. Gardet's discussions on space (p. 91f.) and time (p. 90, 96, 107) which stress the need of regularly "re-making" either, as well as his discussion on numbers (Livre II, Ch. IV) further complement the picture.

Crucial to it all is, as Berr tries to bring out in his preface (p. XVIf.), the strongly and explicitly socialized articulation of all these basic conceptions. The primary manifestation, the primary conceptualization context for the Chinese perception of the universe is society. Society is the prime frame of reference for a thought pattern that, in embracing nothing less than the universe, presents the phenomenal order of things in such a fashion that it appears at the same time as the ideal city. "Tao is not a first cause. It is but an effective total, a centre of responsibility or, again, a responsible 'milieu' " (p. 333).

The immediate practical relevance of all this is indicated by C. K.

Yang (1961). "The influence of Confucianism permeated every fiber of Chinese society through some two thousand years of steady development, with only partial interruption for brief periods. Traditional society cannot be properly understood without giving due consideration to the Confucian doctrine and the institutionalized attitudes of the Confucianists" (p. 244). (See also the contribution by the same author to Fairbank 1957.) To pursue this line of thought is outside the present scope and not strictly necessary to demonstrate the point at hand. The same must be said in regard to the intriguing question, dealt with by such authors as Sarkisyanz (1955) and Levenson (1958), concerning any affinity that could be shown to exist between these ideas and those of communism by which they have allegedly been superseded.

7.1.2. The second case in point should feature slightly below the first in the graphical presentation. (Please correct fig. 8.) Normative ideal order and phenomenal order are largely congruent, but the congruence is subject to some restrictions, mainly due to articulation of concepts. The case is classical India.

Betty Heimann (1930, 6) begins her analysis of the specific nature of classical Indian thought by the statement that "at the beginning of the Indian discipline of thought stands Indian cosmology", that is, the primary conception of the universe. "The point of departure is the phenomenal world given in plenitude of individual manifestations." She tries to show the irrelevance, to such thought, of ideas like divinity as superordinated over phenomenal reality or primacy of spirit over matter or *vice versa*; and she underscores (p. 9) "that all individual manifestations like man, animal, plant, are but equivalent parts of total nature itself", – basic consideration from which ensues "the possibility of changing and interchangeable manifestation of separate parts of nature, the possibility of the doctrine concerning reincarnation of that which has issued from and then reverted back into original matter".

"It is upon this trend of thought – which is concerned not with the separate individual but with the sum total of individuals – that the Indian doctrine of *karma* is based: on the effect of acts belonging to one individual mode of existence upon a new mode of existence of the individual". The author then proceeds to show the implications of these fundamental concepts in regard to ethics. She concludes (p. 11f. *et passim*) that the conditions for a development of ethics (in the current Western sense of a normative frame of reference distinct from phenomenal reality and its basic operational pattern) are absent.

The notion of *karma*, in signalling the complex that is relevant here, must be further elucidated by linking it with two further concepts, *ṛta* and *dharma.*[7] Together, they connote the fundamental conception of cosmos, of given existential order of the universe: ultimate dimension of the ideal city that is relevant both in the macrocosmic and in the microcosmic sense: even though from the viewpoint of human involvement this order, insofar as it connotes a retaliative principle in the sense of Kelsen (1946), appears different in the one case than in the other.[8]

Some further quotations from Betty Heimann's work will substantiate this view. One Arian notion that lies at the root of Indian thought and that, although superseded in the later articulation of the thought pattern, has remained traceable, is the concept of *ṛta.* "... originally (...) the *ṛta* concept connotes that which establishes cosmic order and that thus is useful, harmless, good for all parts of the cosmos (man, animal, plant). (...) secondarily there develops the idea of a particular position of man in regard of *ṛta*" (p. 30). "Being in possesison of his *dharma* of knowledge and judgment concerning his destination and concerning what occurs around him, man faces simultaneously the choice between acting in accordance or not in accordance with *ṛta.*" However, "human ignorance is an offence against his *dharma, i.e.,* world order", and "cosmic damages must be redeemed" (p. 31): again, an "ethics" that by operating in self-propelled cosmic fashion does not correspond to Western notions of ethics.[9]

The author underscores (p. 32) the cultic function of *ṛta* which involves the virtual identification of microcosm and macrocosm, particularly also through the manner in which man becomes integrated in both through the cult. She furthermore points out how the notion of *karma* is the specific Indian elaboration of the basic frame of reference laid down in the concept of *ṛta.* "The subjective individual act – integrated in a fixed *(gesetzmässig)* sequence – becomes an objective necessity, a factor in a continuous, virtually cosmic causal chain that does not even have to become effective in human terms" (p. 33).

With this order of retaliative correlationships (too easily called

[7] On *dharma* see Heimann, 1930, 273.

[8] Betty Heimann says that consideration of the macrocosmic order is primary, of the microcosmic order secondary (p. 29): another way of putting the same thing. Her elaboration on p. 31 however, seems to overstress the difference between macrocosmic, absolutely effective *ṛta,* and microscosmic *ṛta* affected by the human *dharma* of free choice, in that it underrates the importance of the retaliatory principle that is the fundamental mode of operation in either respect.

[9] Comp. Heimann, 1930, 113, 167.

causality) equally operative towards the future as towards the past – which amounts to saying that cosmic order exists regardless of time – the resulting conception of the universe is characterized to a considerable extent by the virtual identity of cosmic and ideal order. As such, it is very close to position III^A (Fig. 9) to be discussed below, to the extent of being almost identical therewith. The difference rests in the difference between myth and normative philosophical system, and besides in the elaboration: a visible order of a compact society there, as against an arrangement of relationships between social-ritual institutions in a diffuse cultural universe in the Indian case. Crucial to the present case, and also distinctive between it and the previously quoted one, is what one could call the interiorization needed to implement the effective identity of ideal and phenomenal orders of existence.

Summarizing the same in the deceptively easy, rationalizing terminology of Radhakamal Mukerjee, "the doctrine of *karma,* as fashioned by the practical religion of the Hindu and the Buddhist, establishes a cosmic moral order itself instead of superimposing upon the human world the working of Divine Justice or the fiat of the Sovereign Ruler of the Universe. So conceived, it holds together life and mind, the present and the past, good and evil, the cosmic and ethical process in an integrated, harmonious whole" (1937, 73).

This is the soil in which have been rooted the successive developments of ritual Indian society: an ideal city not ultimately conditioned by time and space. "Religion is, as it were, the common denominator or the fundamental basis of al the factors which make up Indian life" (Masson-Oursssel c.s., 1934, 61). To pursue these developments in an attempt to bring out the characteristics of Indian society that are relevant to the present focus of interest should be a rewarding task. However, it constitutes a subject by itself that could not be tackled adequately with in the present scope. (Comp. Masson-Ourssel c.s., 1934, Part Two; Prabhu 1963; Mookerji 1954.)

7.1.3. The third case in point to illustrate position II^A (Fig. 8) is traditional Islamic orthodoxy. The presentation, like in the two previous cases, is highly selective and entirely focussed according to present subject. But again it will be attempted to present thought complexes that in addition to being crucial to early Islam have remained determinants of most, not to say all variations on the theme of Islamic civilization.

The writer has on other occasions (1962, 1963/1, 1963/2) analyzed how the prophetic function of the founder of Islam, in conjunction with revelatory data enunciated through him, have remained operative as determinants of Islamic culture and society throughout the history of Islam: Islam being a way of life in the fullest sense of the word. Upon becoming institutionalized, they have been elaborated in such a manner that the revelatory basis of Islam became enumerated under four headings rather than one, viz. scripture proper (codified Qurʾān), revelatory data not codified in scripture (prophetic traditions), analogical expansion of the applicability of data of either kind, and consensus of the Islamic community.

By virtue of the revelatory nature of its main substance, this complex partakes fully in the eternal, divine order of things: witness the much asserted tenet that the Qurʾān is uncreated. By virtue of its immediate relevance to phenomenal reality (the Qurʾān is revealed in "clear Arabic"), the same complex partakes fully again in the phenomenal order of things. The point of coincidence is the community of the faithful, the land of *sharīʿa, i.e.,* of (compliance with) divine law: this latter term being a shorthand expression for the fourfold complex just sketched. This is why L. Gardet is correct in stressing (1954, 22) the "ideal framework of Muslim political society". As he puts it, with a reference to the Christian distinction between the eternal and temporal levels of existence, "Islam has, at its own level, a precise answer concerning the political organization of the Community of the faithful" (*ibid.*).

The Islamic position is therefore singular in that it asserts, in limitative fashion, the virtual identity of ideal city and phenomenal order. Negatively, the limitation operates in consequence of two kinds of distinctions. One is the distinction between Creator and creature; between eternal, absolute order of being and phenomenal, ephemeral order of being; between eternal law absolutely effective and eternal law socially revealed and as such interacting with creation, particularly man.[10] The other is the distinction between the faithful and the unbelievers, between those who have, once and for all, chosen to comply with eternal law and who thus are members of the *umma,* the community, and those who stay outside. Jointly the two distinctions single

[10] The terms of reference in the latter regard are not an-ethical like in the Indian case discussed above, nor emphatically ethical like in Christianity; but in between. One can trace the complex of trespasses – punishment – forgiveness, but rather than ending up in the extreme position of sin as the signal condition of creature, it holds some distant resemblance to the retaliatory principle but personalized. The point, in order to become really clear, must be considered from the angle of the Muslim position in regard to the human free will.

out the one community of the faithful that, by virtue of in principle conforming to revealed law, holds a singular position in all the universe. In a way, it is the universe. Here one finds the positive side of the limitation just mentioned. Socioculturally, albeit not theologically, it amounts to much the same as the chosen people idea. The *umma* is, in all the universe, the community that, in fulfilling the law (or at least explicitly intending to do so), represents the eternal order of things. Phenomenal reality beyond any doubt, it yet is virtually identical with the ideal city. Gardet, following Massignon, speaks in this connection of theocracy (p. 23 *et passim*). The term does not seem to do full justice to the underlying basic concepts; but it may be useful to convey to non-Muslims some of the real meaning of the position.

7.2. Position II^{B} (Fig. 8) in its turn will be further illustrated with some cases in point. They will be presented in an order descending from the centre of the drawing down to the bottom.

7.2.1. The universe that is envisaged in initial Christianity hinges upon the primordial and fundamental cleavage between Creator and creature, between eternal order and phenomenal order. Under the name of original sin, this cleavage reflects in the phenomenal order: as its basic determinant.

The two orders stand fudamentally and categorically opposed, and their opposite nature reflects in various ways in the phenomenal order: within humans as the struggle between good and evil, between humans as the fundamental distinction between those "saved" and those not "saved". At the same time, the intrinsic relatedness of the two orders stands – primarily in the concept of creation and its extension for purposes of historical presentation of fundamentally timeless (mythical) verities: maintenance. Secondarily it also stands in the concept of reconciliation: another basically timeless notion the historical representation of which assumes dual appearances, viz., of unique historical fact and simultaneously of ever-present critical "now". This moment of reconciliation is represented in the mediator figure: exceptional to either order of being in that it fully partakes in both. Just as the former negative instance of the relationship between the two orders reflects in phenomenal being, viz., in the datum of original sin, so this positive instance in its turn reflects in the order of phenomenal being, viz., in the concept of (personal) salvation. Original sin and salvation clearly form a reduplication construct used to render one basic datum (the fundamental human condition) – with a complementary relationship as the modality of togetherness between its two components.

Consequently, their ethical reflection in the phenomenal order of existence is not consistently ethical; but this in a rather different manner than the kind of ethics deriving from the *karma* concept in the classical Indian context. The function of the retaliatory principle, as the mode of operation of the one universal order, is here fulfilled by the principle of redemption, of annihilation of the fundamental difference between two orders of being, resulting in the restauration of one comprehensive universal order. If the former principle is at root an-ethical, the latter is anti-ethical, in that it purposefully eliminates the difference of orders in consequence of which ethics can be envisaged as a relevant matter for one of the two, viz. the phenomenal order. Christian ethics are therefore ethics with a revenge: after all has been said and done they do not really matter: not being more than an inadequate rendering of the basic Christian preoccupation, to become relieved from the order of sin.

Under the conditions, it is obvious that Christianity should show an almost extreme preoccupation with ideas of another world, of an ideal city; at the same time the locus of such ideas remains difficult to pinpoint. The one thing that stands out is the pronouncedly eschatological nature of ideal city ideas in Christianity.

The position becomes clear by comparing it to the Muslim position. As discussed above, the *umma,* in its ideal city aspect, is somehow not quite identifiable with the daily routine of human affairs yet it stands in their midst; besides, there is paradise in the hereafter as a secondary ideal city notion, eschatological but with a point of departure in the historical line of reasoning: projected at the end of time. In Christianity, the same two eschatological perspectives – at the "end" of time and in the midst of time but regardless of it – occur as modalities of its ideal city concept but the accent is the obverse to that in Islam.[11] Heaven, the ideal city at the end of time (more precisely, of personal time, after one's individual death) is envisaged in opposition to that vale of tears which is human life. Besides, there is the notion of the invisible church to represent the ideal city in its other eschatological aspect: in time but regardless of time. As Cullmann points out, the state does not enter into the order of ideal city notions (1957, 18f.).

To further trace the elaboration of the ideal city concept in the

[11] In attributing the dual nature of Christian eschatology to chronological dualism Cullmann (1957, 4,87) seems to overlook the necessarily dual appearance of eschatology in any consciously historical civilization.

Christian context [12] is impossible inasmuch as the one civilization rooted in Christianity that could count as Christian civilization in its own right has had a second root in other than Christian thought and has, for the development of its ideal city concepts, benefited more by the latter than by the former. The one thing that stands out clearly so far as Christianity proper is concerned is that the ideal city is represented, in the framework of the phenomenal city, by the absolute commandment – such as, love thy neighbour like thyself – which besides being the norm, is also the verdict. In the words of De Quervain (1945, 6), ecclesiology is Christology: "The Church is nothing but the next realm of Christ," "Church existence is a life in real responsibilty for state, economy and culture, nay for creation at large" (p. 7).

7.2.2.1. Plato's *Republic* is the next case in point. It may be described, somewhat irreverently, as an exercise in philosophy prompted by wishful thinking: the latter in its turn having been occasioned by sociopolitically determined dissatisfaction. Thus it has two parameters. One is quite pedestrian – the idea of getting away from it all and starting afresh so as to fare better. The (essentially religious) Greek tradition of colonization renders this idea rather more normal and realistic than it might have been elsewhere. The other is purely philosophic – the idea of a speciifc order of reality that represents the true essence of the entities that make up phenomenal reality.

The dialogue sets out from a search for justice *per se* and the discussion on society is introduced as a mere case in point to obtain a firmer grip on the subject. Little doubt is left, however, that prevailing sociopolitical conditions are considered to be quite unsatisfactory in regard to justice (p. 17f., 253; Bk. 8. – Pages referred to are pages of Penguin edition). The effort is, then, in imaginary fashion to found a new society. Meant to serve as paradigm for the discussion on the nature and implementation of justice in society (p. 174ff.), this can but shape up as a full-fledged reconnoitring of the ideal society. In the process, many basic traits of the existing society – some of them definitely conditioning the prevailing mix of justice and injustice – are overlooked, such as its closed community character (pp. 103, 159) and the organic way in which it is commonly conceived (p. 219); also its physical limitations imposed by prevailing organization, communications and productive techniques. Even so, it is intended to start from scratch, like in the founding of a new colony (pp. 161, 102).

[12] Comp. Troeltsch (1931).

Plato's city-state appears to be the ideal city in being "just" (p. 174: wisdom, courage, discipline, justice): his shorthand indication that it is the adequate representation of absolute order. (As shown above, the absolute order is a primarily philosophical category which however does show traces of mythical thought, and this not merely in its alleged need for a myth of origin: p. 160.)

The question can be raised, and is actually raised in the dialogue, whether all this is still real or is it merely ideal: also in the sense of being imaginary (p. 232 et al.). One answer is given; it actually consists of two complementary answers. The first is that in its philosopher-ruler whose insight attains the ideal, absolute order of things, the envisaged ideal state would partake in the ideal order whilst actually existing in the phenomenal order. The second is that education (which, in the light of the Indian example, above, could perhaps be called imposed or induced interiorization) is to bring about the particular conditions necessary to the actual existence of the envisaged society. Together, the answers are more satisfactory in regard to the conceptual problem of how the ideal state can partake in the two orders of existence than in regard to the common-sense problem of whether such a state of affairs can actually occur. The point is that education does not start at zero point, the way the dialogue implies.

In fact, here shows a weakness that appears much more clearly in the elaboration of a number of questions of social structure and division of labour: the amazing disregard for man as a constituent of society with an identity, a continuity of his own. By failing to take these into account and treating of man as an indeterminate entity in a system, Plato has left his ideal city construct unclear at a crucial point. Consequently, in presenting the ideal city as a norm rather than as myth, Plato fails to establish the kind of interaction between norm and phenomenal reality that one might have expected him to take care of – not merely in principle like in the Christian concept of responsibility, but in some form of interaction. This omission is the reason why Plato's ideal city features at the present point in the typology: the signal element is the lack of effective interaction between phenomenal reality and normative image projected in contradistinction therefrom. Lack of interaction that is typical of the gentleman of leisure of Plato's days (Arendt, 1958, 15; comp. further Popper, 1945, vol. 1).

7.2.2.2. Curiously, this typological survey is not the occasion for an adequate rendering of Aristotelian ideas. Here, they are to feature as one variant of the position for which Plato is typical – *n'en déplaise*

Aristotle himself. Another variant is offered by St. Augustine. Between these two variants, one finds the range of ideas for which Platonic thought remains typical.

It is hard to do Aristotle justice in the present connection because the signal work is neither one consistant whole nor is it a rounded-off complex of thought by itself. His *Politics,* as Ellis (1912, vii) remarks, is an elaboration and continuation of his *Ethics*; as Nestle (1934, XXIX, XXXVII) points out, part of it represents the middle period of his life when he has already broken away from docile acceptance of Platonism but is still working on his own ideas, and part represents, or at least presupposes, these ideas in outgrown form. His spiritual growth represents a move away from Platonic idealism, even though not necessarily a complete break with its underlying approach. He is severely critical of Plato here and there, but the reason is perhaps less that he is against the ideal than that he feels the elaboration of the ideal to be carried so much aloft on the wings of imagination that the interaction with the phenomenal order, which after all is a precondition for the ideal order, gets lost. Thus in the matter of Plato's communalism in regard to women (Ellis, 1912, 28) and to property (p. 32).

To Aristotle, the city-state is one demonstration case of his ethics. Accordingly, the state has a purpose (Ellis, 1912, 215f., 225) in terms of happiness for its citizens. How this happiness is envisaged and defined, although it is evidently presupposed, matters curiously little. In the discussion on the state, the implementation of the ethical purpose is effectively superseded by considerations of wealth and power as effective between various social groups and categories. Besides, the view of the city has some organicistic bias (Ellis, 1912, 112f., 214), which remains reasonably well distinct from a preoccupation typical of the culture period, viz. with ranking (Ellis, 1912, 68, 74, 217) of people in various classes and occupational specialisms: preoccupation that culminates in the question of H. Arendt (1952), who is a citizen and who is not, given the rating according to labour vis-à-vis leasure (independence of means).

There is thus a notable difference between Plato's and Aristotle's reasons for writing on politics. To the former it represents a philosopher's exercise in one aspect of ethics, viz. justice, plus a not unwelcome opportunity to uncommittedly criticize the prevailing state of affairs. His ideal state is being speculated upon, not supposing that it would be founded one day. To Aristotle it is much rather a practical "handbook for the legislator" (Ellis, 1912, X): the legislator being an expert whose services are acquired on an *ad hoc* basis by communities feeling a need of them. Thus there is perhaps more of a link between the ideal and the empirical, the earlier and the later sections of the work, in that both angles of approach serve to clarify to the would-be legislator what he must do and how he can try to do it – in the light of *a priori* considerations as well as in the light of actual conditions rapidly surveyed.

This common element, which seems at the same time significant for the nature of the Aristotelian ideal city, shows again in two notable points that continuously recur in the argument. One is the stress upon the mean, upon the necessity to avoid extremes, whether of good or evil, as a precondition to the legislator's work (Ellis, 1912, 92f., 126f., 165, 173, 191). The other is the stress upon the plurality of manifestations, both of actually prevailing political systems and of constitutional frameworks meant to set things right – nowadays one would say, a sense of cultural and political relativity. In both regards, the result of the legislator's activity, the desired constitutional framework, is clearly conceived as standing in effective interaction with the prevailing phenomenal order. Many explain this from one angle only and say that Aristotle is empiricist first and foremost, and does not become unrealistically idealist, particularly not in his mature thinking. Against this view, one is entitled to uphold a complementary appreciation which is based upon an approach from another angle. Then, it appears that the Aristotelian ideal city, in being conceived as standing in interaction with the phenomenal order, and in being presupposed (under the shorthand name of human happiness which is no more than a reference back to the *Ethics;* and also as the constitutional framework that is to result from the effort of the legislator for whom the book is meant) rather than elaborated, is perhaps even more of an ideal, in some respects, than the Platonic one – in keeping with the Aristotelian move away from Platonic metaphysics.

7.2.2.3. Between Aristotle and Saint Augustine lie roughly 750 years of philosophic and religious continuity as well as change. Even so, in certain respects Saint Augustine does not appear to be further distant from Plato than is Aristotle. This refers particularly to the use of the concept of justice (in an un-Roman, ethical rather than legalistic, sense) as the starting point for his ideal city concept. As Sir Ernest Barker points out in his Introduction to the English translation of *The City of God* (Saint Augustine 1945: xii), his conception of four levels of existence (household, city-state or state at large, earth, universe) is geared to the overruling concept of the city of God – with the Platonic term, the city of justice.

This notion, in referring as much to the universe at large as to the (city-)state in particular (note that Saint Augustine wrote when Rome had just been sacked by the Goths, A.D. 410), develops into an ideal type notion of the kind for which Max Weber has at one time become famous. As such, it is the counterpart to the notion of the wordly city – the ideal type concept to indicate the order of unrighteousness. One notes that the latter cannot be identified without more ado with the phenomenal order: the Augustinian construct is by one step more complicated than this. Surely, "be it from us ever to think that we have attained the true happiness whilst we live here" (Vol. II, p. 238). Yet, the construct is further developed so as, on the one hand, not to identify the church with the ideal city (this role is reserved for the so-called invisible church) and, on the other hand, not to reject the phenomenal order lock, stock and barrel: even the phenomenal state "can stand up, on its own basis, with its own justification, to aid the heavenly city" (*ibid.*, I, xix). But in regard to the state in the Roman sense, "St. Augustine shifts the centre of gravity" (p. xxi): the religiously hallowed state which is its own ideal city, is, at the moment of its break-down, said to be "not all, nor indeed is it the half of the matter. There is another and a greater society; and it is towards that society that the whole of creation moves" (p. xxii).

7.2.3. The third sub-type that deserves to be mentioned as a case in point is utopianism, the attitude that was fashionable in 17th and 18th century Europe and of which one can trace the beginning and the successors through a considerable part of the history of European thought. It is to this trend of thought that R. Mucchielli has devoted the main substance of his work (1960) and that has stood model to his phenomenological and historical analysis: in such a manner that also phenomena discussed here under headings III^B and III^D (Fig. 9) are embraced. To sort these various types out, one could perhaps use his own distinction between revolt and revolution. One could then say that the revolt that remains inconsequential, the mere rejection, in terms of ideas, of the existing phenomenal order, stands at the root of utopianism in the sense in which it is to feature here. On the other hand, the effective rejection of the existing (old) order, the revolution, makes for attitudes here distinguished as positions III^B and III^D.

Mucchielli has taken Macchiavelli and Thomas More as two typical instances of utopianism. More's *Utopia* has appeared in 1516, Macchiavelli's *Prince* in 1513. "With Macchiavelli as with More, we recognize the same individual revolt in face of the pityable condition of their people, the same attempt at objective analysis of the situation, with the aim to detect the cause of the social illness they have found to prevail, eventually also the same intention to figure possible means to intervene. The impetus comes from the myth of the ideal city which, in diverse articulation, underlies either's reasoning. But whilst Thomas More ends up in an authentic utopia, Macchiavelli ends up in political realism" (Mucchielli, 1960, 72). Mucchielli ascribes this difference to differences in the articulation of either's pessimism in regard to man and to possible action.

At any rate, what counts to Mucchielli and to present purposes alike, is the basic similarity, which is at the same time the basic utopian conception. "The myth of the ideal city, which has emerged from the intense personal revolt, encounters the pessimism of cool reflection concerning reality" (*ibid.*, 69). In this regard, utopians could easily provide cases in point to the modern interest in alienation: but for the fact that their persistence as upholders of utopia would seem to indicate, in a roundabout fashion, that they would never consider really giving up about society, like many of to-day's victims of alienation do.

Muchielli provides an interesting sequence of moments in thought that, to him, represents the birth of utopian ideas in schematic fashion (*ibid.*, p. 62: (1) individual (humane, non-egoist) revolt against prevailing state of affairs: rationalized *a contrario* in the positive image of the ideal city; (2) realist analysis of prevailing state of affairs; (3) pessimism concerning possibilities of effective intervention, awareness of being inadequate to the challenges; (4) clash between revolt of first and inadequacy awareness of third phase; (5) flight into the imagination, using the myth of the ideal city as the theme for speculative elaboration; (6) elaboration of imaginary ideal order, in literary form, established in contrariety relationship to existing order. "It is the paradigm of an authentic republic, of a healthy state" (p. 60).

This is hardly the occasion to duplicate Mucchielli's and Ruyer's efforts to describe and analyze the thought of the famous utopians. Perhaps one of the most interesting efforts to list them is Mucchielli's attempt to establish a list of what utopians have been against, mentioning names in each case (1960, 101f.). Thus, against oppression, disorder, war, discord, inequality, injustice, corruption, misery and several more.

Rather should one, in an attempt to complement the picture and also in order not to make the impression of passing an unduly favourable moral judgment on this particular attitude, mention the existence of the so-called black utopias. The one that is still relatively fresh in the memory of at least part of the world is nazism. Probably the first to expose it for what it was, and thereby to shock quite a few, but not nearly enough, people out of careless slumber, was Denis de Rougement. His *Journal d'un intellectual en chômage* (1937) and his *Journal d'Allemagne* (1939), to mention these two works only, were dramatic turning points in the orientation of Occidental civilization. The same applies to Hermann Rauschning's *Die Revolution des Nihilismus* (1938). The basic document of this black utopia, rather than Hitler's *Mein Kampf*, has been Alfred Rosenberg's *Der Mythus des 20. Jahrhunderts* (1930). (Note the lapse of time between the first publication of this work and that of the works just quoted.) One short sentence will do to evoke, rather than sum up, what Rosenberg preaches, "Only such myths, with their forms, are alive for which people are prepared to die" (p. 680).

The black utopia, thus exemplified, differs from its counterpart, the utopia in the usual sense of the word, in at least two crucial ways. First, by the manner in which it selects, within the phenomenal order, the elements upon which it will build the ideal order it conceives: no doubt in *a contrario* fashion. As to this first aspect of difference, two remarks are apposite. One notes that as often as not a black utopia is conceived through positive reinterpretation, followed by maximization, of a datum in the existing order that causes apprehension to most non-adherents of the black utopia concerned. For example, racialism or other discriminatory treatment of groups of humans. In the black utopia, a discrimination that used to be relatively inarticulate and carried out with not too clear a conscience will (whether reverted by way of retaliation or in its original direction) become exalted as a prime moral principle, and rendered articulate accordingly. One notes, in the second place, that as such, the growth process of a black utopia presents a most interesting case of deviation in a more or less standard process [13] of overall sociocultural change such as will ensue from a disintegration process having reached an acute stage. The deviation occurs due to selection of the wrong element for the more or less

[13] This statement presupposes certain theoretical ideas on the nature and modalities of social change that cannot be elaborated here. Please refer to my *Society as Process* (The Hague, Mouton, 1962), esp. Annex 2 to the essay "Reality — Solid or Flux?".

standard procedure of reinterpretation-*cum*-maximization. (Instead of reinterpretation, one might say restatement. The idea becomes clearer if translated into German. There, one could speak of *Umwertung* or *Umdeutung*.) Wrong element in the sense that the element chosen, even upon reinterpretation, fails to achieve a fundamental reorientation. Reorientation that, starting from this element, would eventually permeate the entire culture context concerned. Instead, the selected element remains part and parcel of the decrepit old order which, due to the use to which it (the "reinterpreted" element) is put, will persist basically unchanged even though perhaps dressed up in some shoddy disguise.

The second manner in which a black utopia differs from a regular utopia is in its vision as to the relationship between ideal and phenomenal orders. In the former, the vindication of the ideal city is conceived not so much in terms of literary or other presentation, but much rather by moral strength, formal power or actual force: the ideal city must be imposed upon the real order. On the other hand, what is again common between the two types is that both variants of vindication are likely to prove abortive in the long run: given the factual conditions under which utopia and reality are confronted.

These more or less theoretical considerations are borne out in two further examples of black utopias. In mentioning these, it seems useful to caution the reader that the word black, used in indicating this type of utopia, refers to incidental preferences for the symbolic use of the colour black: black shirts, black swastika, and the like. It is never a shorthand indication for race, even where race is involved in the formation of the ideal city concepts concerned.

In the two examples to be mentioned now, the occurrence of negroes as a social entity *per se* in a basically pluralistic society is the point of issue from which the conception of the black utopia starts out. Curiously the traits of the black utopia to which this feature has given rise in the two cases show an inverted parallelism, like a photograph and its negative. The one case is the fear-inspired *apartheid* propounded by a section of white South-Africans; the other case is the exasperation-inspired Black Muslim ideology (or could one call it religion, as they do themselves) propounded by a conspicuous section of the American negroes.

It seems once more unnecessary to produce elaborate quotations. The appalling situation that ensues from *apartheid* as an effective policy stands fully revealed in one simple news item out of all too many (reported in the The Hague newspaper *Het Vaderland,* Sep. 28, 1963). A Durban court has ordered 67 year old, white John de Lange and his 48 year old negro wife Dora Benghoe to divorce. They have been married 25 years, have 6 children and some grandchildren. The minister has confirmed the verdict, which has been passed under the law "against immorality".

On the Black Muslims, a reasonable amount of documentation has gradually been made available. Notwithstanding a display of desire for secrecy and security, which may well be a show of exclusivity, they are quite publicity-minded. Besides, the systematic investigations of Lincoln (1961) and Essien-Udom (1962) provide a fair conspectus. One notes, however, that in regard to ultimate aims, their *apartheid*-in-reversal and racialism remain strikingly elusive. In a TV interview shown on BBC television in August 1963, their then leading spokesman, Malcolm X, stood aloof against sympathetic, yet assiduous and intelligent probing in this very regard. This trait they share with other black utopias.

The point is that black utopias will be fully explicit on certain basic considerations that will determine the features of their ideal city to be. In South-Africa this goes to the extent of detailed legislation and practical provisions to establish extra-urban ghettos for the negroes. However, the ideal city-to-be of which these measures are constituent elements remains largely – that is, for its real profile – beyond the horizon. This would not appear to be a matter of chance. Neither is it, in this case like in other cases, that ideas do not become fully clear in all their dimensions but along with their gradual implementation. Rather it does appear that one encounters here a fundamental characteristic of the black utopia. After all has been said and done it is, for reasons given above, necessarily abortive – not due to coincidences but on account of its intrinsic characteristics.

One way or another, utopia is decidedly the most articulate and elaborate of ideal cities. It is also the one manifestly, though *a contrario,* relevant to the phenomenal state of affairs against which it is upheld. One notes that the *a contrario* congruence operates detail by detail, seriatim, rather than in the way of representing the systematic elaboration of one new positive (as against reinterpreted) maximized principle. Finally, as a norm, it is the least effective.

8. The third type (Fig. 9) occurs in four variants, each representing one of four extreme positions in the typological chart. Considered

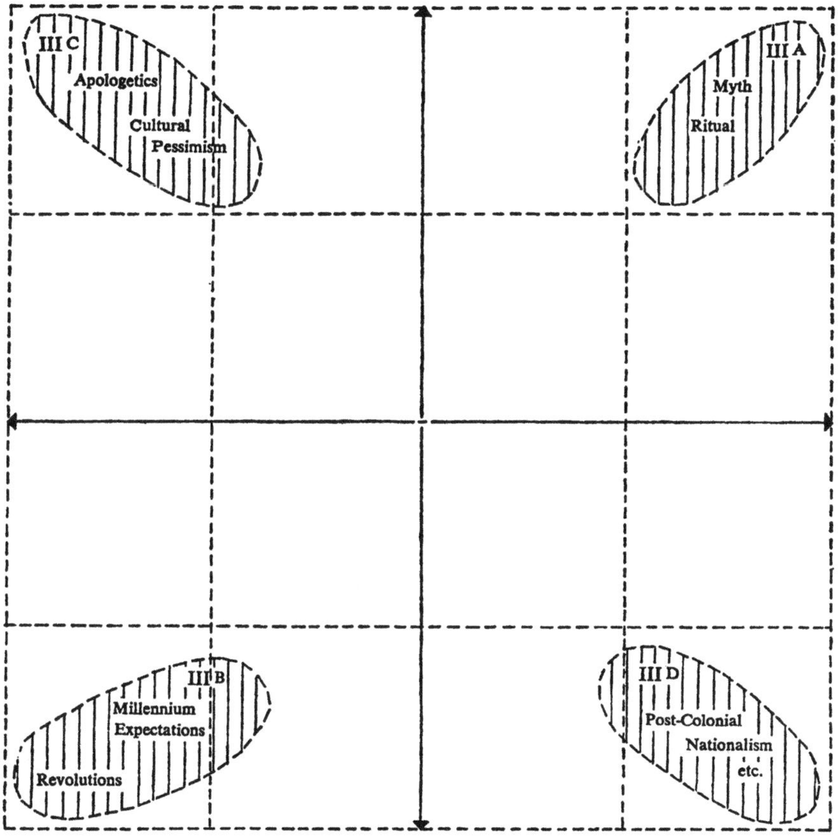

Fig. 9. Idealist approaches, representing attitudes or positions that are likely to prove alternatingly dominant amongst given groups of people, in the course of time.

diagonally they are in a way one another's counterparts: feature recalling the basic complementarities underlying the chart. Considered horizontally, they represent linkages in terms of history of ideas. In the scheme, the four variants seem strikingly apart; in reality they appear to be closely related. (To account for the difference one could perhaps conceive of the chart as representing a global form rather than a flat surface, like in the case of hyperboles in a graph. The point is not very important for present purposes.)

These are all attitudes accounting for the ideal city complex in mythical-idealist fashion. The ideal city is of a higher order than the real city (or community or society). The transcendance that thus characterizes it is either conceived as inherent in phenomenal reality, in interaction with it or pervading it: the two positions of an affirmative attitude in regard to the state of affairs, as shown to the right of the chart; – or alternatively, it is conceived as contrasting with, contradictory even if complementary to, phenomenal reality: the two positions of reaction against the prevailing state of affairs, as shown to the left of the chart. Similarly, the transcendance is either conceived as confirming the phenomenal reality, that is, as its essence, its deeper truth: the two positions at the top of the chart; – or alternatively it is conceived as implying a verdict on phenomenal reality (whether conceived as the present order of things or as the "old" order of things): the two positions at the bottom of the chart.

The use of the composite term idealist-mythical is not meant to suggest that the two component terms are actually or virtually synonymous: far from it. The point is that, from the present viewpoint, the function fulfilled in one context by the mythical complex is fulfilled, *mutatis mutandis*, by the idealist or ideological complex in another.

8.1. Position III^{A} (Fig. 9) represents the myth-and-ritual complex. On another occasion (1958, 3ff.) the writer has attempted to sum up a number of features of what he has there called the closed community which are immediately relevant to present purposes. From its own viewpoint – this being the only one that matters – this community is unique, the (centre of the entire) universe: outside it, there is no really human life. "By the consummation of the myth, in the ritual becoming to the occasion, it ensures a safe passage through life in accordance with life's archetype. This archetype has 'universal' validity (from the point of view of the closed community): by practicing the rite in which it finds its expression the community realizes 'universal' harmony and security" (p. 4).[14]

8.1.1. One example out of very many that can be found in anthro-

[14] This is the occasion to indicate a line of thought that can, alas, not be pursued in the present essay. Myths, besides having the crucial function indicated above, are also a prime form of literature. Considering them in this light, one faces a matter that is most adequately raised under its German name of *Dichtung und Wahrheit*: terms of which "fiction and truth" is a correct yet inadequate translation. A basic trait common to many literary forms is the conscious effort to lie the truth inasmuch as there is no telling it otherwise. As R. Niebuhr puts it with a reference to St. Paul, "as deceivers yet true". The *Iliad*, the *Gilgamesh epic*, the *Nibelungenlied* and so many more are cases in point. But so is the modern novel, that remarkable product of Western civilization. So, in its own particular fashion, is the philosophic treatise – witness the many philosophic writings to which reference is made at many points in the present paper. So, finally, is history writing. Clearly, the concept of the ideal city and the literary way of expressing ideas are closely and profoundly related.

pological literature is the description, summary but directly and strikingly relevant to present purpose, given by Lévi-Strauss (1955, Chs. XXII, XXIII) of a Bororo village called Kejara, on the Rio Vermelho in Brazil. The village is built in a circle around the men's house. "The circular placement of the huts around the men's house is of such great importance for social life and cultic experience that the Salesian Missionaries in the Rio das Garças area soon realized that their most effective means to convert these people was to make them leave their village so as to live in another, where huts stood in parallel rows. Thus the Indios lost the possibility to get their bearings from the main points; once deprived from the plan that served to underpin and substantiate their knowledge, the indigenous people soon forgot sense and meaning of their traditions, as if their social and religious system – we shall see that the two are inseparable – was too complex to do without the scheme that is manifest in the village plan and is constantly renewed in everyday life" (p. 228f.; comp. p. 255). The author then continues to elaborate the details of social structure as manifest in the ground plan of the village, and continues to show how social structure and religious system, the way they meet in the ritual, establish one consistent (view of the) universe (p. 229):

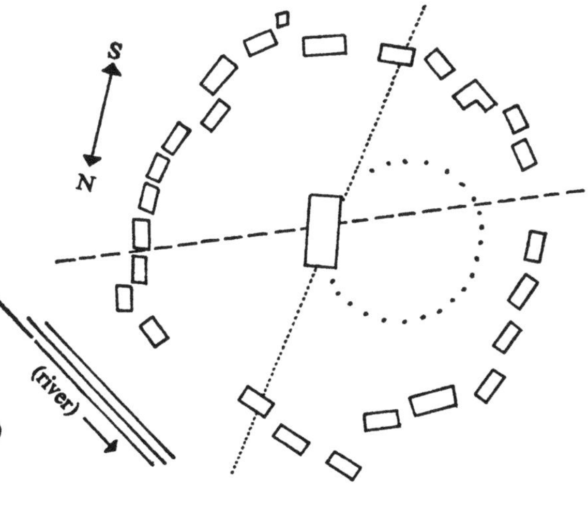

(after drawing in the book)

The narrative becomes the more intriguing thanks to the fact that this is a travelogue and memoir not a monographic research report: the author actually responds to his findings.

8.1.2. From a viewpoint of phenomenology of religion rather than of anthropology, one finds the same matter studied, for the Ngaju

Dayak of Central Kalimantan (formerly Borneo) by H. Schärer (1946, esp. para. 7). This gifted missionary was mainly concerned with the Dayak understanding of divinity or godhead (*Gottesidee*). In establishing the notions concerning godhead he had to pursue their implementation in various directions. Thus, he devotes attention to what he calls, with an adjective that has become fashionable particularly after R. Otto singled it out for a famous analysis (1917): holy people, holy land, holy house. "The holy people, in its origin, existence and social structure, is understandable only in the light of the ideas concerning Divinity" (p. 67). There are holy drawings, kept by the priests, that depict the original village (p. 68). It lies, not in the phenomenal world, but in the superior world; one will eventually return to it forever (p. 69). As to the actual village in which one is born, it is "the largest and most beautiful in all the world and he (the Dayak) would trade it for none other. When he leaves it he takes holy provisions to guarantee his safe return" (*ibid.*). "The world that the water snake carries on its back and that it surrounds with its body is the good, holy country.

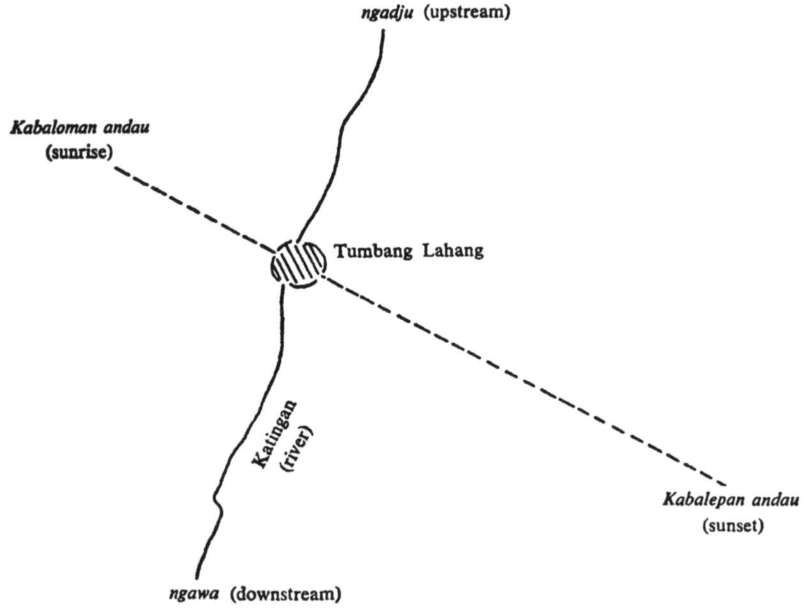

What surrounds the village, *i.e.*, the area that is not delimited and fenced-off by the water snake's body, is foreign, horrible and dreadful land" (p. 70).

Like in the case described by Lévi-Strauss, all this reflects in the village plan. "One's own world is the centre of all worlds, centre of total divine-cosmic order and harmony. This applies also to the village, to which everything said thus far about the holy land is applicable ever since tribal organization has broken down. The village in its turn establishes the social and cosmic totality; in its turn it has the division into two" (p. 74; here follows a discussion on village classification representing, or coinciding with, the order of the universe). The above sketch is copied after Schärer's drawing of the place of the village in the cosmos,

8.2. Diametrically opposite to the position just described is position III^B (Fig. 9). The ideal city stands in the most explicit contrast to the phenomenal city, as an annihilating verdict and as a rallying cry for all those who, far from tolerating the existing order of things any further, are ready either (a) to demolish it and to proceed by establishing the ideal city in the space thus cleared, or (b) to accept the miracle of the new order presenting itself in a fashion not wholly depending upon human action. The sociopolitical revolutionary and the fighter on the barricade are indeed closely related to the man engrossed in millennial or messianic expectation and to the adherent of a cargo cult.

8.2.1. One can actually distinguish a third variation on basically the same theme: the driving urge behind mysticism with its many manifestations – determined according to the institutionalized religion from which it departs and according to many other criteria of distinction. However, the ideal city of the mystic is related to the phenomenal city in a manner so extreme that after all has been said and done it is a relationship of utter unrelatedness. Ethically directed mysticism is not simply contradictory, but in actual fact merely complementary to antinomian mysticism. "The unification of God must happen in the world, man must produce the unification of God by unifying himself; what belongs to man, worldly welfare, understanding, life on this earth, must be ventured on the divine" (Buber, 1948, 82). Thus the mystic stands resolutely outside the present scope: there is no ground for discussing mysticism in this paper.

8.2.2. If one should draw the elliptic form indicating position III^B (Fig. 9) in such a manner that mysticism (not shown in Fig. 9) would actually fall outside the left bottom point of the square, then the variant indicated above (8.2) *sub* (a) should be drawn just inside the square, and the one *sub* (b) a bit further inward, toward the middle of the square. Both however are clearly within the present scope. This subsection is given up to the former of the two, the attitude and world view of the socio-political revolution.

8.2.2.1. Muhammad, the founder of Islam, represents a prophetic revolt against the existing order of Meccan society. "Muhammad's primary experience was an awareness of a disequilibrium and malaise in contemporary society. Correspondingly his aim at first must simply have been to remove this disequilibrium and malaise. This is essentially a negative aim without any positive content. That is to say, the end is not conceived in such a way as to indicate the means by which it is to be realized. Gradually, however, Muhammad became involved in a programme of activity. The beginning of this was the decision to proclaim publicly, at first to sympathetic friends and later to the whole community, the message he believed he had received from God" (Watt, 1961, 50; comp. idem, 1953, 72ff.). "The ideas, then, that are found in Muhammad's preaching contain exaggerations, and to that extent are not wholly true. Their exaggerations, however, do not affect their efficacy as bases for the transformation of Arabian society" (*ibid.*, 52). Being preoccupied with the fact that from this revolution springs Islam, institutionalized religion and way of life of many places and times, Watt has hardly concentrated upon the revolution that makes this integration possible by preceding it and, in so doing, consummating the disintegration of the earlier Meccan way of life.

It is interesting to pursue the historical line that sets in with this revolutionary position. To do so provides a fascinating lesson in the metamorphoses of the ideal city, in the course of time, in a given cultural tradition. After the spiritual estate of Muhammad had been taken over by his successors, institutionalized Islam as a way of life, "la cité islamique", came to represent a position of the type distinguished in position II^{A} (Fig. 8) of the present scheme. But in a state of decay and conscious decline one finds Muslims who deal with the ideals of Islamic society in a fashion that places them in position III^{C} (Fig. 9) to be discussed below. Furthermore, echos of the Islamic ideal city can be traced without difficulty in attitudes such as post-colonial nationalism (position III^{D}, Fig. 9), also to be discussed below.

8.2.2.2. In the present context one can hardly afford not to mention the French revolution.[15] Born out of a demand for reform, the revolution proper enters the scene by the time reforms begin to be seen as demanding the abolition of signal elements of the old established order, by way of a prerequisite to their successful implementation. In the words of Madelin (1924, 41), "he who demands in full earnest its

[15] I have to thank Dr. J. G. Stork-Penning for material on this point as well as for helpful criticisms on this paper as a whole.

reform" (*i.e.*, of the old regime) "runs grave risk of unchaining a revolution". Typical are the reform demands of the *Cahiers du Tiers* which add up to the complete destruction of the existing order. "How delighted should we be if feudalism were annihilated" (*ibid.*). Or, quoting St. Just, "That which constitutes the Republic is the destruction of all that is opposed to it" (*ibid.*, 352; the words date from 1794). Perhaps even more remarkable, coming from a royalist, are the following observations on the occasion of the fall of the Bastille, "the city has severed all its political ties in order to return to (*rentrer dans*) natural law" (*ibid.*, 74).

Complementary to the refutation, which is at once virtual and actual annihilation, of the existing order, is the vision of the ideal order: within easy reach. At the occasion of drafting the declaration of rights, one of the deputies states, "We desire to make a declaration for all people for all times for all countries, and provide the world with an example" (*ibid.*, 84). Signal events in the same perspective are the new administrative division of the country and, much more basic, the introduction of the new calendar. What could be more crucial to any relatively complex civilization than its calendar: what more profound reversal of the prevailing order can occur than the abolition of one calendar and the introduction of another – one that, moreover, is, in a quite curious fashion, more "natural".

8.2.2.3. Even in the briefest listing of examples of the revolutionary attitude, communism cannot be left out. There is no need for a prolonged argument. The following quotations from the *Communist Manifesto* present a world of thought and action in the briefest compass. Although not exemplary for the attitude of the active revolutionary (they stand to the Russian revolution in a relationship not unlike that of Rousseau to the French revolution), they are significant here inasmuch as they contributed to triggering the actual revolution. They are the revolutionary moment *in abstracto*. (And one could not say exactly the same of Rousseau: his book just happened to become the Bible of the revolution.)

"In short, the Communists everywhere support every revolutionary movement against the existing social and political order of things. In all these movements they bring to the front, as the leading question in each, the property question, no matter what its degree of development at the time. Finally, they labour everywhere for the union and agreement of the democratic parties of all countries. The communists disdain to conceal their views and aims. They openly declare that their ends can be attained only by the forcible overthrow of all existing social conditions. Let the ruling classes tremble at a Communistic revolution. The proletarians have nothing to lose but their chains. They have a world to win. Working men of all countries, unite!" (Marx and Engels, 1951, 61).

"When in the course of development, class distinctions have disappeared, and all production has been concentrated in the hands of a vast association of the whole nation, the public power will lose its political character" (p. 51). "In place of the old bourgeois society, with its classes and class antagonisms, we shall have an association, in which the free development of each is the condition for the free development of all" (*ibid.*). The vision of the ideal state and the revolution are preached as the two faces of a coin.

To illustrate the active revolutionary attitude of the Russian revolution, practically any official communist publication on the period abounds with examples. Here is one random example, dealing with events early in 1918. It is taken from a Dutch version of the official party history, in an edition which, it is to be feared, should be immediately identifiable as belonging to the period of history writing before destalinization (*Geschiedenis van de communistische partij der Sowjet-Unie,* 1950). It runs as follows, "In order to achieve the complete liquidation of the remnants of feudalism, of class society and of inequality in any sector of society, decrees were issued concerning the abolition of classes (estates), the abolition of restrictions of church and state and of school and church, the legal equality of women and legal equality of the nationalities of Russia. In a special decree of the Soviet government, known as the Declaration of Rights of the Peoples of Russia, it was determined that the free development of the peoples of Russia and their complete equality before the law are established by law" (*etcetera*).

8.2.2.4. A last signal statement in this series meant to draw the profile of the revolutionary attitude in regard to the ideal state, must be drawn from the recent neo-nationalist revolutions of Asia and Africa. Indonesian President Soekarno, who has never stopped preaching revolution as a continuing process, has once written the following sentences. Referring to the declaration of independence on August 17, 1945, he writes, "What happened there, and subsequently throughout Indonesia, is a revolutionary event. It nullified the Dutch constitution that declared Indonesia to be part of the Dutch Kingdom. This was not a revolutionary event in the juridical sense only, neither exclusively in the political sense. Socially too it was a revolutionary event. (. . .) We have not launched the Indonesian Independence Declaration as a mere declaration: the Indonesian Republic we have not established merely so as to have a state. We ventured these steps by way of spearhead of revolutionary social struggle: as the condition of a struggle to achieve a revolutionary social principle. Our Constitution clearly indicates this revolutionary course, where it says, in its preamble (. . .)" (Soekarno, 1951, 253). (The next part of this statement will be discussed in connection with position III^{D}; see below, 8.4.1.)

8.2.3. An alternative to the revolutionary is offered by him who prepares for the new order to materialize of its own accord. He usually prepares for the event in one way or another, whether spiritually only or also by means of special preparatory measures such as ruining and discarding the implements of the old way of life. There is a wealth of literature on the subject; which is not surprising, given the ubiquitous nature of the phenomena concerned.

Lanternari (1960) has recently brought a wealth of material together in a highly commendable conspectus. The more recent volume edited by Sylvia L. Thrupp (1962) can be seen as a collection of additions and further methodological elaboration to this work. Wallace's attempt at the theory on revitalization movements (1956) can be further referred to as a viable effort at elucidation. Compare also Linton (1943).

Lanternari's focus of interest is well summed up by the reviewer of the *Times Literary Supplement* (Sep. 29, 1961) who writes, "For the most elementary alternatives of the most primitive spiritual reaction to conquest – war to expel the whites (as with the cults which inspired the risings of Tecumseh and Sitting Bull), are soon found inadequate. Cults may arise which seek chiefly to re-establish social equilibrium after the catastrophe of white irruption, seeking to bring under control the vices introduced by the whites (as among the Iroquois and in the Peyote Cult, both with a marked bias against alcoholism) or the proliferation of indigenous evils, such as witchcraft. Very soon, however, the problem of absorbing and controlling the new western power and technology emerges. The Melanesian cargo cults look forward to the arrival of the millennial cargo of the white man's goods, which will enable the native to get rid of the white man and to enter the age of abundance. What is more, certain ideas and institutions of the white man – forerunners of western revolutionary ideology – become the nucleus of salvation within the oppressors' civilization." To the responsible Western civil servant in whose area it may occur, a cargo cult or other millennial expectation movement presents a threat for the consequences it may entail: gruesome consequences dramatized in Nicholas Monsarrat's *The Tribe That Lost its Head*. There are, however, cases where it ends in a more creative, though even then perhaps not all too promising fashion (Mead 1956, esp. p. 214ff., 412ff.).

8.3. Position III^{C} (Fig. 9) represents the variant of the mythical-idealistic approach to the ideal city that sees ideal and real city as emphatically distinct, to the extent of conceiving the two as of virtually different orders. Its response to the discrepancy that it finds between them is not a creative response to something experienced as a challenge. Rather it is a position of the mind that abstains from any action involving ideal and phenomenal worlds at once.

It suffices for present purposes to indicate two signal attitudes, as two different examples of this type.

8.3.1. Apologetics is an art of argumentation. It confronts a complex of valued ideas with a complex of actually existing conditions in such a manner that their mutual relevance is proven. Inasmuch as the ideas concerned are supposedly sacrosanct and inviolable, phenomenal reality bears the brunt of the impact of the reasoning involved. Sometimes this does not appear too clearly, due to an all too woolly appearance of the apologetic argument. Sometimes phenomenal reality does not become effectively involved in the argument, insofar as the latter assumes the form of an aprioristic statement against real or imaginary opponents who argue that the ideas in question are irrelevant to the actual conditions in question. But sometimes there is an overt *tour de force* in regard to phenomenal reality, either by representing it in a fashion that disregards its true or full appearance, or by simply redefining it.

One example out of many that can illustrate the latter position can be found in a pamphlet by Sayyed Abulala Maudoodi, leader of Pakistan's fundamentalist Jamaʿat-e-Islami. It is available in English translation (1947). After some argument concerning economic problems he decides "the economic problem of man to be no more than this: with a view to sustain and advance human civilisation, how should economic distribution be arranged so as to keep all men supplied with the necessities of life and see that every individual in a society is provided with opportunities adequate to the development of his personality and attainment of the highest possible perfection according to his capacity and aptitude?" (p. 12). This surely is a far cry from the customary understanding of economic issues in an underdeveloped area of predominantly Muslim culture (with its formal prohibition of rent). But is is good enough for the argument that follows. It first ascribes the evils in the economic system to the natural selfishness of man when it exceeds the limits of moderation (p. 19). After rejecting what is presented as the communist, fascist and such-like solutions, the argument culminates in propounding the Islamic solution. What is presented as such is an eclectic and sketchy presentation of Islamic ethics and regulations concerning "religious taxes". The conclusion is that "the economic system of Islam becomes a great blessing for humanity. Studied with an unbiassed mind, free from those prejudices against Islam which have been inherited as an evil legacy from past history and undeterred by the blaze of modern social systems, I hope and trust that this Islamic system will satisfy every reasonable and truth-seeking person as the most useful, correct and rational method for the economic welfare of humanity" (p. 60f.).

Writings of this nature are by no means exceptional nowadays. The style seems particularly attractive to authors whose training and education are partly along traditional Islamic lines, partly in terms of Western schooling. Pamphlets, books and periodicals of this nature are found throughout the world of Islam and even where Muslims live in Western countries.[16] One notes that as against the phenomenal world, the ideal world is maintained in an entirely peculiar manner. Islam as a way of life, as the norm and therefore as the ideal city in the sense used here is referred to categorically. In most cases it is presupposed rather than propounded, let alone elaborated. The usually summary and eclectic presentation offered actually avoids establishing the relevance, detail by detail, of the normative system to reality. Thus apologists often appear to be more concerned to stake a position for themselves with which and by which they can be identified – perhaps also in terms of public opinion formation, social control and political power – prior to (and sometimes rather than) using this position as their point of departure for the realization of creative interaction between established norm (ideal city) and phenomenal reality (actual society). The danger of apologetics is that it falls short of doing justice not to phenomenal reality only but just as much to the ideal norm behind which it entrenches itself.

The position is neatly summed up by I. M. Husaini (1956, 40f.) in

[16] Some further examples, by way of random choice: Khallaf 1930, Hamka (undated), Ahmad 1950. Also the periodical *The Islamic Review* (Woking, England).

regard to Hassan al-Banna and his Muslim Brethern: "he says, 'In this world there is no system which supplies the awakening nation with the organization, foundation, feelings, and sympathies as does the system of Islam'. He also states in this letter that the Koran gathers together all the sciences of the world in one verse. He urged them to follow the Koran. This was also the message of Muhammad ʿAbdu. The letter reads that the rules of Islam as they pertain to the individual, the family or the nation, its government, its people, or the bonds tying nations together, is at once comprehensive and precise and gives preference to the general welfare; and that they are the most complete and beneficial rules known to man. He believes that Islam has an all-inclusive meaning, organizing all the affairs of life and interpreting all matters and that it lays down exact and precise rules for them."

8.3.2. Cultural pessimism represents an attitude in terms of ideal city conceptions that is not identical with apologetics yet so close to it that discussion under the same heading in the present scheme is warranted. The ideal city is anxiously considered as paradise lost: leaving no or virtually no hope for phenomenal reality.

8.3.2.1. Thus, to begin with another example from the wide realm of Islamic civilization, the nostalgic interpretation of *colonisabilité* by the Algerian Muslim author Malek Bennabi (1954, 23f.). Speaking of the fashion to discuss Islamic civilization in terms of decadence (fashion that has once caused an international symposium to meet and devote scholarly attention to the matter: see Brunschvig et Von Grunebaum 1957), he writes, "On this matter the Muslim world is in particular need of clear ideas to guide its present effort at renaissance. This is why it is of primary importance to take into account the distant causes that have determined its decadence. The Muslim world underwent its first rupture at the battle of Siffin, in 37 A.H., because so soon after its birth, it already contained an inner contradiction: the spirit of *jahiliya* in conflict against the Qurʾanic spirit. (. . .) From this first rupture onward (. . .) the Muslim world has lost its original equilibrium, even if the Muslim has been able to remain deeply attached to the spiritual order inherent in his believing soul." And further, "the development that has become known under the name of Muslim civilization was nothing but adjustment on the part of Islamic doctrine to the factual conditions after Siffin" (p. 56). Certainly, Bennabi propounds that the return to the initial purity of "before Siffin" should enable present-day Muslims to make a fresh start and achieve the renaissance, the achievement of the ideal state, for which they are yearning.

But he hardly shows how they could. In this regard, the reformer in him, who today sees a vocation for Islam and wants it to be fulfilled, is at cross purposes with the cultural pessimist in him, who yesterday ended up bewailing the lost ideal.

8.3.2.2. Basically the same position is ascribed by Geyl to Huizinga, in a recent article that attempts to draw some more or less definitive conclusions from the total oeuvre of the famous historian. He feels that "all through he was (...) closely tied to his present – through opposition, through disapproval, through constantly saying 'no', through near-despair" (p. 259). These "irritations and (...) glooms were the counterpart of positive and fervent feelings of attachment, always, of course, to the social order or forms and institutions fitting in with his conception of culture, or which he saw as requisite for it, to the exclusion (...) of much that had at least as valid a claim to recognition. I mean the forces of dynamism and of change, in which he could see nothing but destruction" (p. 259-260).[17]

Above (*sub* 6), it was already pointed out that in mentioning Huizinga as one case in point, one actually refers to an entire literature of which he, in fact, is not even one of the most marked exponents. This is the, in itself notably varied, type of writings that occupies itself with the culture crisis of the West – or by whatever other name the thing may be indicated. Spengler, Ortega y Gasset, René Guénon, Toynbee, as well as James Burnham and Hannah Arendt, and ever so many others, are signal names in the connection. Some of these, as the ones just listed, are first rate thinkers each of his own kind; others, who nevertheless have managed to produce best sellers from time to time, do perhaps not rank quite that high in perspicacity. There is some comfort in the realization that contemporary Western civilization is by no means unique in having produced such an output of culture pessimism. Other periods in Western civilization have had their own version of cultured people's despair about their own culture; and so have other civilizations in various periods. The present analysis of ideal city concepts is one way of tracing some of this despair in its changeable manifestations.

8.4. Position III^{D} (Fig. 9) represents that variant of the mythical-

[17] For further clarification one adds that the culture pessimist represents another position than the utopian (II^{B}, Fig. 8). If it be true that the effort of both has an element of *frappez toujours*, it represents a literary exercise with the one and a moral obligation with the other. If it be true that both tickle their readers, the result, if they are successful, is praise in the one case and apprehension in the other.

idealist approach that, in bringing ideal city and phenomenal reality close together, conceives of their interaction not in terms of the virtual identity between myth and phenomenal reality but in terms of the implementation of a model, the realization of a programme, the achievement of a lofty goal.

8.4.1. One signal phenomenon of this order is found in preambles to constitutions and similar documents such as declarations of independence. The battle of minds or the revolution is over, and this kind of documents signal the moment at which the realization of the ideal city is about to begin. Declarations of general principles,[18] they provide the bearings for the main ground plan according to which the city, the new society, is to be built – nay, has already begun to be built. (Comp. W. F. Prins, 1950.)

The passage from Soekarno's *Sarina* (1951, 253f.) quoted above (8.2.2.4.) continues as follows, quoting from the preamble of the Constitution of the Indonesian Republic, "In order to promote public well-being, to develop the life of the nation and to help realize a world order based upon freedom, lasting peace and social justice, Indonesian national freedom is laid down in a Constitutional Law of the Indonesian State, which state is shaped as an Indonesian Republic: with people's sovereignty, founded upon the overlordship of God Almighty, just and cultured humaneness, Indonesian unity, and a public spirit guided by wisdom in the consultations of the representation; realizing, moreover, social justice for the entire Indonesian people." [19]

8.4.2. Another signal phenomenon in the same order, but somewhat nearer to phenomenal reality still and supposedly functioning as directive in its regard, is the development plan. The development plan that is conceived for limited periods such as five years features normally as one out of a series of such plans. Thus, it represents one link in a chain and the latter's purpose is avowedly to reach goals which, if they are hopefully meant to partake in the phenomenal order of things, are to at least the same degree representative of attempts to achieve the ideal city.

One may trace elements of this underlying trend of thought even in the sober and concise wordings of the Second Indian Five Year Plan

[18] A particularly interesting aspect of these general principles is the question of their meaning if, as often happens in cases of post-colonial independence, they represent a cross-cultural impact. (See Van Nieuwenhuijze, 1958, 208.)

[19] A related phenomenon is constituted by the platforms of political parties. On account of the many variants and complications, an analysis of this case in point is less rewarding for present purposes.

(Govt. of India, 1956, 21), "Significant as the achievements of the first plan have been, it is apparent that they have to be regarded as no more than a beginning. The task is not merely one of reaching any fixed or static point, such as doubling of living standards, but of generating a dynamism in the economy which will lift it to continually higher levels of material well-being and of intellectual and cultural achievements." "For several plan-periods to come, it is on the mobilisation of the effort rather than on the gains and returns arising therefrom that attention has to be concentrated. These gains and returns are important, but more important is perhaps the satisfaction that a community gets from attempting a worthwhile task which gives it a chance to bend its energies to productive and socially useful purposes. The 'costs' of development, viewed in this light, are a reward in themselves. There is no doubt that given a right approach to problems of development, including social policy and institutional change, a community can draw upon the latent energies within itself to an extent which ensures development at rates much larger than nice calculations of costs and returns or inputs and outputs may sometimes suggest" (*ibid.*). Furthermore, "the pattern of development and the lines along which economic activity is to be directed must from the start be related to the basic objectives which society has in view. The task before an underdeveloped country is not merely to get better results within the existing framework of economic and social institutions but to mould and refashion these so that they contribute effectively to the realisation of wider and deeper social values" (p. 22).

8.4.3. A third kind of signal phenomena of still the same order is established by the specific nature and function of political leadership under conditions such as follow immediately upon a successful revolution or upon the achievement of independent statehood.

The crisis in self-identification which then marks the existence of the nation (or other social unit) concerned is signalled by a convergence in sociocultural (also economic, political) functions. The ideal city which, so long as the revolution or struggle for independence lasted, seemed, or was believed, to be just around the corner, appears suddenly to be remote. What first may have seemed to become the moment of its realization proves merely the fulfilment of certain basic conditions, that allows an effective effort towards its realization henceforth.

Under such conditions, political leadership towards the ultimate goal, the ideal city, is effective inasmuch as in its actual mode of operation it is felt to foreshadow, to anticipate the full manifestation of the ideal city. It brings the goal nearer by giving it a name, then perhaps the beginnings of a shape. In the political leader, the followers detect the ideal city to come; he is their bridge towards it across the abyss of the present. For those who have disowned their past in anticipation of their future ideal city, the present holds little besides the political leader and his acts.

8.4.3.1. Writes Jamal ʿAbd-el-Nasr, president of UAR-Egypt, "We live in a society that has not yet taken form. It is still fluid and agitated and has not yet settled down or taken a stabilized shape. It is in the process of an evolution, striving to catch up with those other nations that have preceded us on the road" (1956, 68). "The way is that of political and economic freedom, and our role is that of a guardian, no more and no less. And our guardianship is only for a specified and limited time" (p. 71). "To repeat and emphasize, our task is but to mark the way, to shepherd the strays so that they may start and remain together on the right path; to point out to those who may go off after the mirage the emptiness of their delusions. From the start I knew that this would never be an easy task, and I realised beforehand that it would be accomplished at the expense of popularity. We had to be blunt, outspoken, armed with reason whenever we addressed the people" (p. 72). "Our revolution must be sustained by our having the courage to embark on whatever is deemed necessary, no matter what loss of popularity and applause and cheers such action may cost us. Otherwise we will have failed in our trust as leaders of the revolution" (p. 74). "It is our cherished duty, our moment of historical responsibility, to thus bring our people at long last together, and weld them in unity for the future – the future of Egypt – strong and free" (p. 77ff.).

8.4.3.2. It is of course seldom that one can expect the self-view of the leader to be so clearly and factually expressed in terms immediately relevant to present purposes. But the same vein of apperception is inherent in such documents as the famous Speech of Kemal Atatürk (Kemal, 1929) or the publications of Nkrumah. From the latter, the following quotation bears witness (1961, 210). "Thus the State exists in the image of the people and operates confidently in the service of the people. When the Party and the Government take your interests to heart and fulfil their responsibilities towards you, it is your duty in return to help the Party and the Government by your devotion to our cause and your steadfast loyalty to all our ideals. Our record for the past ten years stands before the country as an example of what we can do in the future. I am simply asking you on my own behalf and on behalf of the Convention People's Party, under all circumstances and at all costs to stand united with us (. . .)." Elsewhere, he speaks of "the new life which we are trying to create in this part of Africa" (p. 244). Curiously, the ideal that is foremost in Nkrumah's mind as a leader becomes much more articulate in regard to Africa as a whole then in regard to Ghana as a political entity by itself; witness his idea of 'the African personality' (Ch. 15) and related concepts.

8.4.3.3. Note that colonial and post-colonial nationalism, in order to be adequately understood in its fundamental urges, should be seen in comparison to (although not identified with) European utopianism rather than to European nationalism. If nationalism has provided the modality of fixation to a given abstract-and-concrete phenomenon, the nation, it has been utopianism that has stood model for most of the inner urges and stirrings. But more important than to compare either way is to realize that what is primarily at stake is the urge toward self-realization: the recapture of awareness as regards one's own society, envisaged in the perspective of its proper ideal image, along with the ensuing effort towards realization of one's own ideal city.

9. The fourth type here to be distinguished (Fig. 10) represents what is in a way the most realistic attitude towards the ideal city: that of him who is about to build it. Its place in the typological chart is astride the horizontal line, inasmuch as its attitude towards the existing city is naturaly bound to imply rejection and confirmation simultaneously. Even if one is lucky enough to build in entirely open space, there are geographical and other (*e.g.*, economic) determinants to consider, and plans do not come out of the blue either. But it is decidedly to the right of the vertical line that runs through the middle of the chart, inasmuch as it represents an unequivocally constructive, creative, organizing not entropy-promoting stand in regard to reality. This is further brought out by the consideration that whenever this type, the builder's attitude and effort, happens to be related to the ideological movement or the social revolution, the relationship occurs after the movement has gained control of things and after the social revolution (whether industrial revolution or political revolution) has become the new social order.

9.1. Sir Patrick Geddes is the father of modern town and country planning. The term town planning becomes clear in the following statement, "Too many (. . .) are plainly accustomed to think of town planning as an art of compass and rule, a matter to be worked out, between engineers and architects almost alone, and for their town

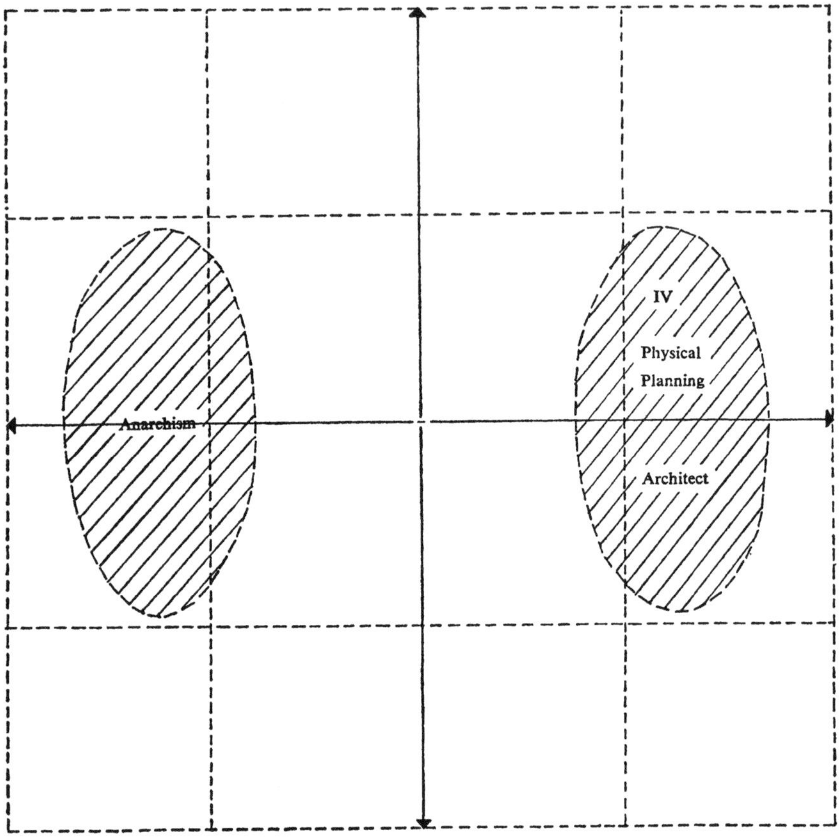

Fig. 10. The builder's approach – spatial organization: the new city, urban planning, town and country planning.

councils. But the true town plan, the only one worth having, is the outcome and flower of the whole civilisation of a community and of an age. While starting from its fundamentals, of port and road, of market and depot; and from its essentials too, of family dwellings worthy to be permanent and hereditary homes, it develops onwards to the supreme organs of the city life – its acropolis and forum, its cloister and cathedral" (1949, 71). His motive views are avowedly utopian, but in what is meant to be a particular way. As against the paleotechnic order and its corollary, Kakotopia, he proposes Eutopia as the goal to be achieved by neotechnic order (p. 38f.). His concern with technology is less marked than these terms suggest: it remains instrumental and as such is mostly implied in his reasoning rather than brought to the fore. In fact, he has a quarrel with technology insofar as it becomes technocratic. Thus, for example, his harsh view of Hauss-mann's boulevards of Paris (p. 105f.): he condemns them for their social consequences. Indeed one of the focal points of his thought is his concern with society at large; a concern that has a more than out-wardly esthetic component. He wants to cleanse the city (p. 51), more particularly its slums (p. 55) – note that slum cleansing is not necessarily slum clearing – and has in fact cleansed his own chosen slum neighbourhood in Edinburgh. He is also concerned with the beauty of cities (p. 48f., *et passim*).

But the esthete in him remains visible where he speaks as the comprehensive social scientist who, with the survey (Ch. 15, 16) as his main tool, tries to gear a number of approaches to his purpose by putting them under one common denominator: civics. An object to be taught and a field of studies (Ch. 14), civics is primarily a mode of creatively grasping the spirit of cities (Ch. 17). It thus brings out the City with capital C. "At this stage the City becomes again reviewed as a whole, as he who understands a town-plan sees all the town as from an aeroplane. All our activities – industrial and commercial, hygienic and educational, legal and political, cultural and what not – become seen in relation to one another, as so many aspects and analyses of the city's life. To make this life more healthy and more effective, the unrelated individual activities with which we have been too long content are found insufficient; we need fuller co-ordination and harmony of them, like that of the instruments of the orchestra, of the actors in the drama. We expect this of soldiers in the field, of workers and organizers in the factory, of assistants and partners in the business. Is it not for lack of this orchestration, of this harmonious organisation, upon the larger civic stage which our town-plans so clearly reveal, that our cities, full of detailed efficiencies of many kinds, are still so far from satisfying us as collectively efficient? The time, then, is ripe; the place is every city; each needs its Civic Survey and Exhibition, its Civic Study and Laboratory" (p. 87).

If this were not enough indication of an ideal city viewpoint, one notes two further points.

First, Geddes' reliance on the exhibition – that extreme form of idealization – as a signal operation in the context of the effort he advocates in others and is constantly making himself.

Secondly, an attitude to the city as a living entity that betrays at once the evolutionistic biologist he was by training and the esthete he must have been by birth. "Every considerable city, in short, seeks to complete itself. It no longer contentedly accepts provincial inferiority; it finds itself with the means, and increasingly with the will, to develop its own civilization within and not merely draw it from without" (p. 90).

9.2. From the man who pursues his ideal city by means of organization of social and physical space to the architect is a most natural step. In fact, the difference is a matter of degree of involvement in technological know-how and activity. After Geddes, whose main contribution – in to-day's view – is perhaps more in his organic, synoptic approach to the valley as a unit of habitation than in regard to the city, one thinks of Frank Lloyd Wright, for many years an institution in American architecture. His contribution was certainly in architectural creations in the first place. But as the founder and leader of a school of architecture and as the author of various works on the future function and development of architecture, he holds a place of his own in the ranks of those concerned with the ideal city.

In fact he has tried to build it, if only in a small scale model. Its name is Broadacres; it is to be, in the terms of one of his book titles, *The Living City* (1958). Like Geddes, he tries to achieve a kind of universalistic approach to things and he coins a terminology and a style of his own for the purpose. Like Geddes, he rejects the existing order of human togetherness, as habitat and as type of community, and maximizes some of the salient new achievements of this order (mainly technical achievements) as stepping stones towards the new order he envisages. The new city (p. 99ff.) becomes possible thanks to newly achieved mobility for people and decentralization possibilities for everything else. The old order, symbolized by such phenomena as rent which tend to render man a slave, is rejected outright, rather than conceived as the point of departure for building the new order: here the anhistorical American in Wright (in the sense of Cl. Lévi-Strauss, 1956, Ch. XI) is diametrically opposed to the historically embedded European in Geddes.

Further symbolic terms used for the new order are freedom and organic (p. 160) architecture, also beauty; but insofar as the terms become clear at all, there is still more (architect's) estheticism than in Geddes and rather less of the social critic or revolutionary. If revolt there is, it is the revolt of the town planner who, in the words of W. H. Whyte Jr. (*Fortune*, 1957), does not like towns. Terms like feudal, to describe and condemn the old order, come too easily to ring true. Indeed, Broadacres is subtopia writ large. "When democracy triumphs and builds the great new city, no man will live as a servile or savage animal; holing in or trapped in some cubicle on an upended extension of some narrow street. Withstanding all this passing danger to him – the free man will again live free: the human biped which the best of him always dreamed of being! Life and love as noble leaders of our brave social experiment" (p. 110).

"Charm and comfort" and "pleasure in work and leisure" are indicated as signal elements of the new life; "And imagine man-units so arranged and integrated that every citizen may choose any form of production, distribution, self-improvement, enjoyment, within the radius of, say, ten to forty minutes of his own home – all now available to him by means of private car of plane, helicopter, or some other form of fast public conveyance; factories in which to make his living. Such integrated distribution of living all related to ground – this composes the new city embracing this entire country: the Broadacre City of tomorrow. The city becomes the nation. When every man, woman and child may be born to put his feet on his own acres and every unborn child finds his acre waiting for him when he is born – then democracy will have been realized" (p. 127, comp. p. 132, 140, 151).

"Thus we shall see entirely new forms for living; and see a finer, more secure family life. We will see integrity of means to ends, the individual we have been prophesying and promoting we shall see taking effect in more intelligently civilized constructions. Men would build from the heart as well as by the head: build either by hand or by machine, but build always from the good ground upward and from the inside outward to comprehend time, place and man" (p. 152). There are some thoughts on the community that is to be born out of all this architectural conditioning; they are not really far afield from current ideas and ideals common to all subtopias. This, however, is to be subtopia with no megalopolis to flee away from. If revolution there is, it is implicit, not actually part of Wright's thought. The rapturing vision of the ideal city is enough to carry him away from the sordid real state of affairs of megalopolis.

9.3. There is no more dramatic occasion for planner or architect to be engrossed in the ideal city than when his effort meets the effective desire of an institutionalized social revolution to manifest itself in the city. There have been such happy coincidences in the past, like when Akbar decided to build Fathpur-Sikri (1569). Nowadays, a number of occasions attract attention. Most notable are three new capitals, each the expression of new sociopolitical awareness and each extremely daring and revolutionary viewpoint. Brasilia, Islamabad, Chandigarh. The cases are the more intriguing inasmuch as the architects whose names have been conspicuously attached to the realization of the ideal city that will render the new (national) social philosophy visible are, in two cases out of the three, complete aliens to the ideas in question. The work in India was made famous by attaching a Frenchman's name to it, that in Pakistan became the city of a Greek: leading architects both, but known for quite pronounced and effectively publicized architectural views of the ideal city that are by no means rooted in the spiritual context where they have had to work at these assignments.

9.3.1. Le Corbusier (pseydonym, no initials) was perhaps one of the most, and most sumptuously publicized architects: his complete works from 1910 to 1957 have appeared in 6 impressive volumes. In his opening statement (vol. 1910-29, 7e éd., p. 11) he says, "what appeals to me is the magnificent phenomena of architecture – I mean the intellectual quality inherent in its organisation – which, for the creative mind, constitutes a system capable of expressing, not a mere reflection of one's personal predilections, but a microcosmic synthesis of contemporary history"; with a foot note added, "Architecture to me means construction achieved by a triumph of the intellect." In November 1922 he contributes to the Salon d'Automne in Paris an eye-catching proposition to improve Paris, under the slogan, "Paris expects from this period: the rescue of its threatened life, the protection of its beautiful past, the magnificent and mighty manifestation of 20th century spirit" (*ibid.*, p. 38; see also vol. 1934-38, 6e éd., p. 140ff.). Of these proposals he says at a later occasion (*ibid.*, p. 110) that they represent a doctrine of urbanism. (Comp. his 1963 booklet.)

Although the conceptual background remains hazy, one feels somewhat nearer to it in the following (vol. 1929-34, 6e éd., p. 119), "The unrest of the present time on final analysis goes back to the housing problem. Out of the element house, follows the conception town, to town belongs flat land – and one arrives at a reconstruction of the whole land through the new organization of agriculture and town building" (*sic* !). Inasmuch as he feels that a new epoch is beginning, the civilization of technique (p. 120), he maintains that "industry must conquer building" (p. 122). The result should be a new type of apartment (p. 121).

At one point, he claims that a guiding principle for city reorganization must be the attempt to use the twenty-four hour daily rhythm to proper advantage (vol. 1934-38, 6e éd., p. 20ff.), and then argues that "The new city is compact and occupies less space. It has no suburbs" (p. 21). Like Wright he feels that mobilization and mechanical speed require a complete revision of ideas concerning the city (vol. 1936-46, 4e éd., p. 147). But again, his argument (if not his drawings) halts at the point where he claims, "When mechanistic civilization will have understood that harmonious living, *i.e.*, habitat, work, care of body and mind, circulation, must be the very object of its stock taking of the past, then will 'urbanism' put order to the social facts, and architecture contribute the daily well-being" (p. 149). He wants to gear all the new technological and artistic possibilities to this aim (p. 151).

One would like to know more about these ideas when seen in direct connection with the Chandigarh project. But in this regard merely factual information, wrapped in some rhetoric, is provided (Vol. 1946-52, 3e éd., p. 114ff.; vol. 1952-57, 2e éd., p. 51ff.).

Can one say that in writings like those of Wright and Corbusier just reviewed, their notion or vision of the ideal city becomes clear? No indeed. In fact, it remains elusive. Their words somehow do not convey ultimate meaning. Hence perhaps their struggle with neologisms – which in their turn do not produce the desired clarity. The reason may be that in order to become clear, their vision of the ideal city must not be verbalized but materialized: in a town plan, in buildings. The society for which the town will provide the spatial and material framework remains to be filled in, and all the architect can do is evoke it in idealizing and vague words. But then, could a poet or a musician who fully conveys his meaning in his proper medium – poetry or music – paraphrase the same meanings in prose?

9.3.2. The architect of Islamabad is Dr. C. A. Doxiadis, whose office at the top floor of his Athens Technological Institute, high on mount Lycabettos, offers him as much a synoptic view of the Athenian universe as Geddes obtained of Edinburgh in the camera obscura of his Outlook Tower. (Something in Wright's Taliesin responds in its turn to the same basic urge: the synoptic view, the exhibition hall with models, charts, graphs, maps and photographs, and the private educational establishment.) Here again an approach to an architectural recreation of natural space into ideal city, that sets out from a sweeping theoretical view that operates in almost universal categories and that establishes its own terminology in the process. An approach that, in spite of the theorizing that accompanies it, remains exemplary of the artist rather than of the social scientist, and where research and drawing room personnel operate in a manner that, modern efficient organization notwithstanding, can in some respects be compared to the group of pupils and helpers that used to surround the master painter or sculptor of other days.

From amongst the many writings of Doxiadis, his paper presented to the Second European Congress of the Regional Science Association (1962) may be chosen as characteristic for present purposes. First, by way of explanation of his term Ekistics: "irrespective of how we look at space, if we concentrate our attention on the settlements which are the nodal points for the total, we deal with one subject and this subject is space, no matter if for reasons of systematic analysis of its elements we call it space dedicated to production, transportation or other living purposes. By the way: in every scale of space, small or big, once it is settled by man we have the homogeneous part, the lines of communication and the nodal points. This is true of a house or farm, of a factory or of a metropolis, of the whole earth. What makes the space of interest to Ekistics (and not only to natural geography) is the fact that it is settled or ecumene" (p. 16). And furthermore, "In order to succeed in Ekistics we should understand the importance which the unity of purpose has for all projects, programs and policies. We cannot be successful unless we understand that all efforts should be equally satisfied from the economic, the social, the political, the administrative, the technical and the cultural-aesthetic points of view" (p. 60).

His vision of the phenomenal city appears in the following, "From the village we move to the city, to the metropolis, to the megalopolis and gradually we are witnessing the formation of several types of megalopolis in many parts of the world. In a century or two from now all these are going to be interconnected in a continuous network. In the meantime this is happening already in several parts of the world. The network of Ecumenopolis is very different from the networks of the human settlements of the past. It does not consist any more of small isolated settlements within the vast countryside. On the contrary it is creating continuous networks of major nodal points and many radial branches which are covering the whole earth like a great octopus, and gradually, because of the interconnection of the branches, they are surrounding the countryside within the elongated parts of Ecumenopolis" (p. 66).

As against such a prospect, which to Doxiadis does not seem to appear as too threatening, the vision of the ideal city becomes manifest mainly in implicit fashion; as a challenge to the daring planner. "If

Ekistics or those disciplines which preceded it (town planning etc.) have been able to overlook the real substance of their task, that is the necessity to look at settlements only as the nodal points of all functions within broader areas or regions (and this very fact has led to the decline of the disciplines related to human settlements and their inability to solve their problems, in the last century especially), this will no longer be possible, as the human settlements are conquering the earth and merging themselves with the other areas on a much greater and much more apparent scale than in the past" (p. 75). What Doxiadis envisages is a super-problem, to which his super-science of Ekistics (p. 35, 37) is to be the adequate response. Between the two stands an implicit vision of the super-city as a complex of communication nexus. His big dream is to master the problem of terrestrial space: ultimate size of the ideal city. But whether this ecumenopolis of the graphs and charts will still belong to the human dimension does not become clear.

9.3.3. A few impressions from visits paid in early 1964 to Chandigarh and Islamabad may round off these observations based on the writings of the architects concerned. The cities lie in the same kind of location and in the same kind of climate; in the wide plain just at the foot of a chain of hills linking up to the highest mountain system on earth. Both are administrative centres, with industry relegated, in the one case, to the outer fringe and, in the other case, to the adjacent and pre-existent town of Rawalpindi. Both are planned to make use of purposeful large-scale landscaping, including an artificial lake playing a crucial role in shaping the city. Here the comparability ends. Chandigarh has a checkerboard pattern with introvert blocs surrounded by roadways; the administrative centre, with the four famous buildings by Le Corbusier, stands conspicuously distant from the town proper. Significantly, the governor's residence is in a very good but much less conspicuous location on the edge of town; this not necessarily by Le Corbusier's decree. Islamabad on the other hand is based on the typical Doxiadis axial design, in this case resolutely restricted to basically one axis, thanks to clever use of the hills, a specially created natural park and the existing town of Rawalpindi. The nodal point from which the dynamic plan of Islamabad emanates in ever widening concentric circle-segments or rims, is the administrative centre, with the presidential palace standing on its own hillock in the most conspicuous position. Thus, symbolically, the two conceptions end up as significantly different. Islamabad could perhaps symbolize Pakistan as the administrative development state, not without an authoritarian or paternalistic element. Chandigarh is less clear for its symbolism, unless one were to attach meaning to the relative remoteness of the administrative centre and to the modesty of the residential quarters as compared to it. Even so, in regard to either plan and either city one finds occasion to wonder in which way it would effectively, manifestly symbolize the new Punjab (or India) and the new Pakistan, respectively.

9.3.4. These are not the only cases in point. On the other side of the Iron Curtain a well-entrenched ideology, victorious in a revolution, has engendered conscious efforts to implement the ideal city. The architects are perhaps less likely to feature as prima donnas than in some other situations, but much of the setting remains comparable. Not so the actual city that is built so as to reflect the prevailing social ideals.

J. C. Fisher (1962) has contributed a most interesting study on the subject, mainly based on Polish, Yugoslav and such-like efforts at reshaping existing cities. "Socialist planning desires to create a new form, a new pattern for the city. The overall goals of socialist planning are: (1) to correct the ills inherited from the era of capitalism; (2) to develop a new pattern for the city which will indicate clearly the inherent unity of the people, the classlessness of the society. The spatial manifestation of the new socialist relation among men is at the same time to conform to the stated ideals of each country's national revolution. According to socialist theory, the city is the core of the ideal communist community, a centralized, highly standardized and uniform type of social organization" (p. 205). "The plan of spacing in a socialist city must provide all necessary clues as to the socialist relationships of the respective parts of the city and should find its expression in the common link and not in the division of buildings into isolated islands. There is no sharp distinction or structural division among the various parts or components; the quality of the residential areas and their social composition is to be everywhere the same" (*ibid.*). The author distinguishes four basic operational principles of socialist urban planning, viz., standardization, restrictions in size of town, city centre is civic not retail centre, divisions into neighbourhood units. "Throughout Eastern Europe, thousands of fields and meadows have been transformed into urban complexes to house the rapidly growing working class. These genuinely new cities – constructed on previously non-urbanized sites – should represent the utopia of urban planning in socialist design" (p. 207).

9.4. If from the position as exemplified in Sir Patrick Geddes one line runs to the architect of the New Towns, another line runs to the regional planner and the town and country planner. As Glikson (1955, 8) puts it, "The objective of Regional Planning was defined by Benton MacKaye as 'the cultivation of habitability' of the regions of human settlement." "Habitability, therefore, is a state of balanced relationship which has to be achieved by man by careful selection, adaptation, planning and action. In the establishment of this balance lie the tasks of Regional Planning: Its aims are not the realization of any specific objective, but the regulation of relationship between human and environmental factors" (p. 9). As compared to Geddes and Wright, the statements made here are more circumspect even if they echo the same terms. "Regional planning sees a change of outlook as a prerequisite for the continued existence of civilized life. It is true that Regional Planning does not presume to be a new social or political movement. In contrast to such movements Regional Planning considers the relation of man to his environment and resources at least as equal in importance to the relationship of men among each other. Regional Planning, therefore, works on a program of actual works to be done in order to improve man's relation to his environment. This practical subject – the stable settlement of man, the habitability of regions of settlement, transcends the limits of contemporary categories of mechanistic thought. The idea motivating this kind of planning strives at the coordination of human needs and the special characteristics of man's material environment, not by blind utilitarian behaviour, but organically – *i.e.* in a manner by which man's specific ways of life foster also life in its environment – in a manner by which man finds the place of a cooperating partner in nature's land communities and cycles" (p. 12).

9.5. The typology offered here is – it should be unnecessary to underscore the point – meant for ad hoc purposes. It is a classificatory device meant to meet the requirements of the present subject. A good way to demonstrate its limitations, even for such a restrictive purpose, is to raise the question how to deal with anarchism within the context of our scheme. Anarchism is a complex phenomenon, as recently shown again in G. Woodcock's volume on the subject (1963). Considering it, and in particular the individual anarchist, under one angle, one may satisfactorily range it under heading III C of the present scheme, along with culture pessimists and the like. Considering it, and in particular anarchist movements, from another angle, one would much rather locate it in position II B, along with the utopians – be it not without some hesitation as to possible placement in position III B, along with the revolutionaries.

In consequence of such difficulties one faces a dilemma of either saving the unity of the concept anarchism by providing, say in Fig. 10, an area on the left hand side of the scheme, symmetrically opposite area IV indicated there; – or alternatively, sacrificing the unity of the concept anarchism on behalf of a perhaps slightly more logical placement for some of the isolated strands of thought that constitute this highly complex – and often self-contradictory – phenomenon. For present purposes it seems plausible to opt for the former alternative, and place anarchism *in toto* in the left hand space of Fig. 10 facing position IV.

9.5.1. There is occasion to wonder what examples to offer in trying to illustrate the characteristic attitude for which anarchism stands. Insofar as anarchism is that phenomenon which one finds in individual anarchists of thought and action who somehow achieved a degree of conspicuousness and of fame, there is no doubt that quotations from written work risk to miss the essential point. Woodcock has good reason to point out, in regard to several of the anarchists he describes (Godwin, Stirner, Proudhon, Bakunin, Kropotkin), that writing, this main art of communication, was not what they excelled in: Tolstoy featuring almost as an exception to the rule. Insofar, on the other hand, as anarchism is, willy nilly, a movement or organization, there can hardly be any question about the signal value of communicated messages. Even so, one's inclination remains to look for the signal statement amongst the standard bearers.

One such signal statement, and a very famous one, comes from Proudhon and has been duly reproduced by Woodcock (1962, 105). "If I were asked to answer the question: 'What is slavery?' and I should answer in one word, 'Murder!', my meaning would be understood at once. No further argument would be needed to show that the power to take from a man his thought, his will, his personality, is a power of life and death, and that to enslave a man is to kill him. Why then, to this other question: 'What is property?' may I not likewise answer, 'Theft?'."

And from Kropotkin this one (*ibid.,* 191), "Man is not a being whose exclusive purpose in life is eating, drinking, and providing a shelter for himself. As soon as his material wants are satisfied, other needs, which, generally speaking, may be described as of an artistic nature, will thrust themselves forward. These needs are of the greatest variety, they vary in each and every individual, and the more society is civilized, the more will individuality be developed, and the more will desires be varied."

From Tolstoy, finally, this (*ibid.*, p. 209), "The modern state is nothing but a conspiracy to exploit, but most of all to demoralize its citizens. (. . .) I understand moral and religious laws, not compulsory for everyone, but leading forward and promising a more harmonious future; I feel the laws of art, which always bring happiness. But political laws seem to me such prodigious lies, that I fail to see how one among them can be better or worse than any of the others. (. . .) Henceforth I shall never serve any government anywhere."

10. The distributive treatment that can be accorded to signal types of the ideal city complex in the context of a typology as here developed necessarily demands a conclusion in one comprehensive approach. What, then, is the ideal city?

Even the comprehensive answer has a number of distinct aspects. *Logically,* the ideal city concept results from conceptual duplication on the basis of phenomenal reality, the duplicate being envisaged as standing in a complementarity relationship to the latter. *Psychologically,* it is projection, or *Gestalt,* starting out from and simultaneously exceeding phenomenal reality; by liberating man from phenomenal reality to an extent, it allows him to deal therewith in his capacity as the maker of culture. *Sociologically,* it is two (complementary) things at once: on the one hand it is one manifestation of social identity and as such the relatively stable frame of reference for phenomenal society; on the other hand, in being distinct from phenomenal society, it is one manifestation of the all-pervading dynamic principle in culture and society – the built-in destabilizer that is inherent in any social structure as such. *Economically* (one hesitates to add) it is that which economic theory has lost sight of by pervertingly defining it as equilibrium and which it is now trying to retrieve under such inadequate names as self-sustained growth. *Historically,* it is the meaning, the sense behind – and sometimes opposing – the appearance of things: engendering events and inspiring deeds. *Religiously* it is that which unites the this-wordly and the other-worldly orders – to Christians, at once invisible church and heaven; to Muslims, at once *umma* and paradise –: absolute reality that is simultneously beyond and in phenomenal reality. In the shorthand rendering of a last summing-up, it signals man, the human mind, as the crucial agent in culture and society.

10.1. Thus, ideal city and phenomenal city are commensurate: in a very particular manner. Ideal city notions are likely to exist referring to any size or type of institutionalized and organized human society, whether village, city, state or empire. This mutuality of relevance is clearly instanced where Fustel de Coulanges (1864, Livre III Ch. I) notes that "the religious idea and human society were thus to grow in scope simultaneously" (with reference to the scope of Roman gods). Putting the same in other words, one finds that the ideal city complex constitutes a most impressive paradigm of cultural relativity.

All this has its consequences for the manner in which one may hope to add meat to the typological skeleton produced above.

10.2. Muchielli (1960, Livre II) and also Mannheim (1936, Ch. IV), and in his own fashion Ruyer (1950), have attempted to sketch a more or less comprehensive image of the ideal city as a concept or complex per se. Mucchielli gives the procedure applied an adequate name in calling it the phenomenology of the ideal city. Indeed, these efforts amount to a listing, in some kind of order, of signal appearances, both of a general nature and in terms of main variations on this general pattern. There is relatively little danger in such attempts inasmuch as the authors have taken the framework of Western civilization for granted in their dealings with the subject. Thus, they have avoided the awkward problem of plurality of phenomenologies as necessitated by plurality of given culture contexts. On the other hand this circumstance inevitably reduces the theoretical value of their – by no means negligible – achievement. Insight gained is not really on the general level, as some might inadvertently suppose; much rather it is on a level that one may (with a term that, curiously, refers back to Mannheim once more) indicate as the middle range.

At this level one finds that for the Western culture context there are indeed a number of traits common to the wealth of varieties in articulation of the ideal city concept. The headings used by Mucchielli are significant in the connection. One is "harmonie et transparance": cosmos in the full sense of the word, an orderly, integral universe. The second is "la communauté humaine universelle": the humane in its fullest possible development, both qualitatively and in regard to spatio-temporal extension. The third is "la république": a social order not subject to constraint exerted upon any of its component elements. The last is: "la loi morale": the achievement of ethical perfection – of what is called *Heil* in German.

In the connection, adjectives acquire signal value. In fact, they become crucial inasmuch as it easily becomes their role to feature as

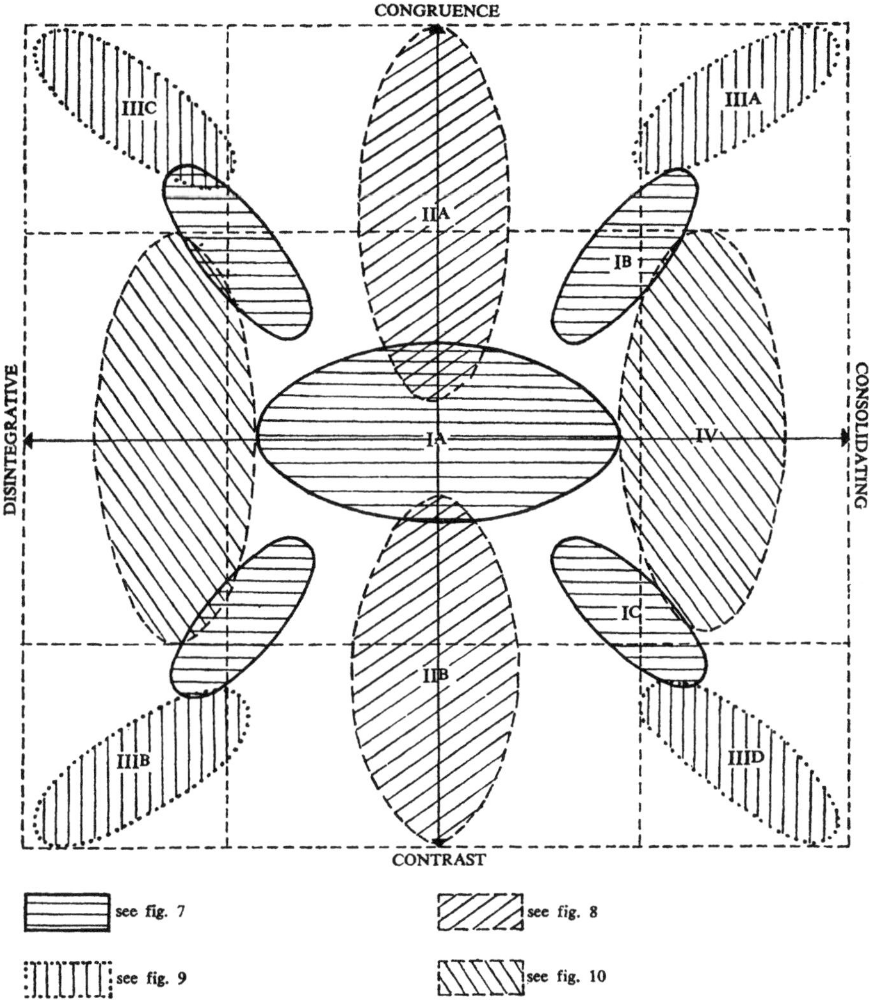

Fig. 11. Composite picture of the four main positions distinguished in the typology.

rallying points for opinion formation, as crucial elements of an argument or of a harangue. Certainly, they are, in the process, overloaded with meaning: and this quite purposefully. To say that under such circumstances an adjective becomes a shorthand term for an entire complex of ideas is very far indeed from doing justice to what is really the matter. Adjectives like "free" or "natural" have on many occasions achieved the power of a key to a new universe or, what has at times been the same, of a percussion-cap in a bomb. Similarly, "just" and "happy" have at times been enriched with semi-metaphysical dimensions not usually belonging within the confines of their meaning.

10.3. The phenomenologist, whether of one culture or of many cultures, is not the only one who is in a position to contribute to the understanding of the ideal city complex. Each of the viewpoints or disciplines listed in the opening paragraph of this section could be expected to offer its own contribution.

10.3.1. One such contribution, on the part of sociology, has been ventured by Mannheim (1936). His preoccupation, on the occasion, was with the sociology of knowledge. But starting out from his considerations and findings one could equally well concentrate on the impact of the ideal city, in a sociology not of knowledge but of society as process, the fundamental dynamism of which is instanced in the tension between ideas and reality.

10.3.2. To study the societal impact of the ideal city is a proposition comparable to other impact studies aimed at understanding of social dynamics or of those particular manifestations of (certain kinds of) dynamics which are currently fashionable under such names as change and development. One culture element is then distinguished, *ad hoc,* from the total culture context to which it pertains and an analytically relevant relationship construed between the two. But to the purpose of such studies, he who chooses the ideal city has an advantage over him who takes up such things as technology or organization. The latter are no more integrally part and parcel of total culture than the former; yet, as appears from the typology above, the articulation of culture tends to allow considerable distinctness, in the daily routine of society, to the former whereas it accords virtually none to the latter.

10.3.3. The study of impact presupposes, in the case of the ideal city, a considerable amount of factual information; more, in fact, than is readily available, yet by no means impossible to collect and organize. This information may well be of a purely descriptive nature and in terms of statics, provided it accounts fully for cultural relativity. It is in the latter regard that much of Mannheim's preoccupation (whilst dealing with ideology and utopia) with sociology of knowledge, becomes ploughed under.

This knowledge about what may be called, with a variation upon William James' title, the varieties of metasocial experience, would issue from comparing entire culture contexts (the phenomenal identity of which is presupposed).

It would mainly refer to varieties, more in particular,

(a) in existing ideal city complexes: phenomenologically considered, with reference to a pluricultural rather than to a unicultural context (the pluricultural expansion of Mucchielli's phenomenology);

(b) in relevance relationships prevailing between ideal city and corresponding phenomenal city (further elaboration of typology above);

(c) in congruence, distributively considered, between ideal city and corresponding phenomenal city:

- (i) in regard to point(s) or focus(i) of coincidence or alternatively of contrast beween the two;
- (ii) in regard to social locus(i), or agent(s), of the confrontation that bring out this coincidence or contrast;
- (iii) in regard to rationalization(s) underlying the act of thus confronting the two: whether
  - (I) with reference to phenomenal reality – summing it up
    - either in criticism
    - or in support;
  - (II) with reference to the ideal city – summing it up
    - either as distinct from and as a verdict on,
    - or in corroboration of phenomenal reality.

10.3.4. The study of impact proper, setting out from the availability of relevant case material along the lines just sketched, will concentrate upon shifts, as instances of the dynamic aspect of things. In the light of the scheme just proposed, it is likely that these will be envisaged as shifts either in the relevance relationship or in the (distributive) congruence between ideal city and phenomenal reality.

Given this dual approach, the shifts proper can be handled in more than one manner; they also have to, inasmuch as one would really look for the understanding that can be hoped to ensue from the cumulative result of these different manners.

(A) The shifts just indicated will be attributable to particular shifts in (component elements of) the ideal city complex concerned or in the corresponding phenomenal reality:

(1) monoculturally;

(2) cross-culturally.

Either way, the noticeable phenomena can be summed up as abnormal tension in the (congruence and/or relevance) relationship between ideal and phenomenal order. In cross-cultural cases, shifts in elements can be further analyzed as penetration followed by adoption, of elements from alien contexts, whether of ideal city or of phenomenal order nature, into the ideal or phenomenal order concerned. The importance of studies of this nature appears in many writings on acculturation. Besides, it is clearly shown – to quote one recent example – by A. Hourani (1962).

(B) The modalities of the impact ensuing from shifts thus analyzed will have to be studied under two headings, each too well known, as analytic procedures, to demand comment or elaboration here:

(1) the time lag involved in the impact;
(2) the operational pattern of the impact (chain, with or without metamorphoses; spread like oil-stain; catalyzer effect; etc.).

The analytic procedure under both main headings will benefit by application of breaking-down techniques as suggested in the previous sub-section under (c).

10.3.5. There is nowadays an inclination, amongst certain groups of people who are concerned with development in such a manner as to desire it, to believe that nothing is as important to actual development as ideology. Ideology in the connection is a shorthand expression for the impact of a (new or changed) ideal city concept that will direct people's minds and actions in such a manner that development will naturally and necessarily ensue.

This clearly assumes a causal chain that can scarcely be expected to occur very often, if at any time. Simultaneously, in clearly staying within the domain of wishful thinking (except for totalitarian regimes), the belief just signalled seems to convey a sufficient dose of remaining awareness as to the interaction, of an emphatically complex nature, that prevails everywhere between ideal city and phenomenal order. Change or guided development, regardless of whether they start at the ideal city end or in the phenomenal order, will always demand the time-consuming frictional process of mutual readjustment, between ideal and phenomenal order as well as between component elements in either order. Most of these processes, moreover, occur not as one causal sequence but in complex simultaneity through time.

10.4. One more conclusion remains to be drawn – a thoroughly practical one. The contemporary age has its own particular mood – it is a time of doing. Social action prevails over social theory; and the main reason for the existence of social theory that public opinion is prepared to recognize is its capability of subservience to social action. This applies to underdeveloped areas for purposes of development, as much as to developed areas for purposes of sustained progress.

Social action can, by and large, be identified as the attempt to implement some kind of ideal city. It is the effort to realize some idea that as yet does not appear as (sufficiently) manifest in the phenomenal reality concerned. This being so, the effectivity, the chances of success in social action stand to be influenced, in a crucial manner, by the relationship that happens to prevail between the ideal and phenomenal cities concerned. This consideration, moreover, is bound to assume added critical importance any time when ideal city and phenomenal city involved happen to be cross-culturally connected rather than related in the more "natural" fashion of belonging to one and the same culture context. This happens any time when social action strives to implement ideas or ideals exogenous to the culture context in which it operates.

It is in this regard that the desirability manifests itself for close inspection of the relationship between ideal city and phenomenal city as a matter deserving attention in its own right. As such it offers one argument in favour of the heuristic value of the typology developed in this paper.

REFERENCES

Ahmad, Z. A., *Dasar-dasar ekonomi dalam Islam* (Djakarta, Sinar Ilmu, 1950).
Arendt, H., *The Human Condition* (Garden City, N.Y., Doubleday Anchor, 1958).
Baratz, J., *A Village by the Jordan; The Story of Degania* (London, Harvill, 1954).
Bennabi, M., *Vocation de l'Islam* (Paris, Ed. du Seuil, 1954).
Beveridge, W. H., *Social Insurance and Allied Services: The Beveridge Report in Brief* (London, H.M.S.O., 1942).
Brunschvig, R., et Von Grunebaum, G. E., c.s., *Classicisme et déclin culturel dans l'histoire de l'Islam* (Paris, Besson-Chantemerle, 1957).
Buber, M., *Hasidism* (New York, Philos. Libr., 1948).
Corbusier, Le, *Œuvre complète* (Zürich, Girschberger, 1910-1957).
—, *Manière de penser l'urbanisme* (Genève, Gouthier, 1963). First ed. Paris, 1946.
Coulanges, Fustel de, *La cité antique* (Paris, Hachette, 1864).
Cullman, O., *The State in the New Testament* (London, SMC Press, 1957).
Doxiadis, C. A., *Ekistics and Regional Science* (Athens, Doxiadis Ass., 1962). (Doc. R-GA 265; mimeo.)
—, *Architecture in Transition* (London, Hutchinson, 1963).
Ellis, W. (transl.), *The Politics of Aristotle, or A Treatise on Government* (London, Dent, Everyman's Library, 1912). (1947 reprint used here.)
Essien-Udom, E. U., *Black Nationalism, A Search for an Identity in America* (Chicago, U.P., 1962).
Fairbank, J. K., *Chinese Thought and Institutions* (Chicago, U.P., 1957).
Fisher, J. C., "Planning the City of Socialist Man", *Journal of the Amer. Inst. of Planners*, 1962. (Used here: extract occurring in *Ekistics*, Athens, 15/89, Apr. 1963, p. 205ff.)

Fortune, Editors of, *The Exploding Metropolis* (Garden City, N.Y., Doubleday, 1957).

Freyer, H., *Theorie des gegenwärtigen Zeitalters* (Stuttgart, Deutsche Verlags-Anstalt, 1963).

Friedmann, G., *Fin du peuple juif?* (Paris, Gallimard, 1965).

Fung Yu-lan, *The Spirit of Chinese Philosophy* (London, Kegan Paul, Trench, Trubner, 1947).

Gardet, L., *La cité islamique, Vie sociale et politique* (Paris, Vrin, 1954).

Geddes, P., *Cities in Evolution* (London, Williams & Norgate, 1949).

Geschiedenis van de Communistische Partij der Sowjet-Unie (Bolsjewiki) (Amsterdam, Pegasus, 1950).

Geyl, P., "Huizinga as Accuser of his Age", *History and Theory*, II/3: 231-262 (The Hague, Mouton & Co., 1963).

Glikson, A., *Regional Planning and Development* (Leiden, Sijthoff, 1955).

Government of India Planning Commission, *Second Five Year Plan* (New Delhi, 1956).

Granet, M., *La pensée chinoise* (Paris, Renaissance du Livre, 1934).

Halpern, B., *The Idea of the Jewish State* (Cambridge Mass., Harvard, U.P., 1961).

Hamka, *Islam dan Demokrasi* (Boekittingi, Tjerdas, undated; ±1947).

Heimann, B., *Studien zur Eigenart indischen Denkens* (Tübingen, Mohr, 1930).

Hourani, A., *Arabic Thought in the Liberal Age; 1789-1939* (London, Oxford U.P., 1962).

Husaini, I. M., *The Moslem Brethren; The Greatest of Modern Islamic Movements* (Beirut, Khayat, 1956).

Kallen, H. M., *Utopians at Bay* (New York, Th. Herzl Found., 1958).

Kelsen, H., *Society and Nature: A Sociological Inquiry* (London, Kegan Paul, Trench, Trubner, 1946).

Kemal, M., *A Speech Delivered by Ghazi Mustapha Kemal, President of the Turkish Republic, October 1927* (Leipzig, 1929).

Khallaf, A. W., *Al-Siyāsatu 'l-sharī'a.* (Used here: transl. H. Muhammad, *Dasar*^{2} *Negara Islam*, Jogja, Tiro, 1947).

Lanternari, V., *Movimenti religiosi di liberta e di salvezza dei popoli oppressi.* (With bibliography) (Milano, Feltrinelli, 1960) (Transl.: *Religious Movements of the Oppressed*, New York, Mentor Books, 1965).

Leibnitz, G. W., *Principes de la nature et de la grace fondés en raison, Principes de la Philosophie ou Monadologie.* Ed. A. Robinet (Paris, P.U.F., 1954).

Levenson, J. R., *Confucian China and Its Modern Fate: The Problem of Intellectual Continuity* (Berkeley, Univ. of Calif. Press, 1958).

Lévi-Strauss, Cl., *Tristes tropiques* (Paris, Plon, 1955).

Lincoln, C. E., *The Black Muslims in America* (Boston, Beacon, 1961).

Linton, R., "Nativistic Movements", *Amer. Anthrop.* 45 (1943), p. 230-40.

Madelin, L., *The French Revolution* (London, Heinemann, 1924).

Mannheim, K., *Ideology and Utopia, An Introduction to the Sociology of Knowledge* (London, Routledge Kegan Paul, 1936).

Marx, K. and Engels, F., *Selected Works in Two Volumes* (Moscow, Foreign Lang. Publ. House, 1951).

Masson-Ourssel, P., de Willman-Grabonska, H., Stern, P., *Ancient India and Indian Civilization* (London, Routledge Kegan Paul, 1934).

Maudoodi, Sd. A., *Economic Problem of Man and Its Islamic Solution* (Lahore, Maktaba Jama'at-e-Islami, 1947). (Used here, 1955 reprint.)

Maunier, R., *The Sociology of Colonies*, 2 vols. (London, Routledge Kegan Paul, 1949).
McDonald, J., *Rousseau and the French Revolution 1762-91* (London Univ., Athlone Press, 1966).
Mead, M., *New Lives for Old; Cultural Transformation — Manus, 1928-1953* (London, Gollancz, 1956).
Mookerji, R. K., *The Fundamental Unity of India* (Bombay, Bharatiya Vidya Bhavan, 1954).
Mucchielli, R., *Le mythe de la cité idéale* (Paris, P.U.F., 1960).
Mukerjee, R., *Theory and Art of Mysticism* (London, Longmans, 1937).
Mumford, L., *The City in History: Its Origins, Its Transformations, and Its Prospects* (London, Secker & Warburg, 1961).
Nasser, Gamal Abdul, *Egypt's Liberation. The Philosophy of the Revolution* (Washington D.C., Publ. Aff. Press, 1956).
Nestle, W. (transl.), *Aristoteles Hauptwerke* (Leipzig, Kröner, 1934 (Taschenb. 129).
Nieuwenhuijze, C. A. O. van, *Aspects of Islam in Post-Colonial Indonesia* (The Hague, Van Hoeve, 1958).
—, "The Prophetic Function in Islam: An Analytic Approach I", *Correspondance d'Orient: Etudes* 1-2, p. 119-138 (Bruxelles, Centre p. l'ét. des probl. du monde musulm. contemp., 1962).
—, idem II, ibidem 2-3 (1963). (1)
—, "The Qur'an as a Factor in the Islamic Way of Life", *Der Islam*, 38/3, p. 215-257 (Berlin, De Gruyter, 1963). (2)
—, *Cross-Cultural Studies* (The Hague, Mouton, 1963). (3)
Nisbet, R. A., *Community and Power* (formerly "*The Quest for Community*") New York, Oxford U.P., Galaxy Books, 1962).
Nkrumah, K., *I Speak of Freedom: A Statement of African Ideology* (London, Heinemann, 1961).
Otto, R., *Das Heilige. Über das Irrationale in der Idee des Göttlichen und sein Verhältnis zum Rationalen* (München, Beck, 1917). (Ed. used here, 23.-25. Aufl. 1936.)
Plancommissie van de Partij van de Arbeid, *De Weg naar Vrijheid: Een Socialistisch Perspektief* (Amsterdam, Arbeiderspers, 1961).
Plato (transl. H. D. P. Lee), *The Republic* (Harmondsworth, Penguin Books, 1955). (1962 ed. used here.)
Popper, K. R., *The Open Society and Its Enemies*, 2 vols. (London, Routledge Kegan Paul, 1945).
Prabhu, P. H., *Hindu Social Organization*, 4th ed. (Bombay, Popular, 1963).
Prins, W. F., "De plechtige aanhef van constituties", *Orientatie*, 29 (Djakarta, 1950).
Quervain, A. de, *Kirche, Volk, Staat: Ethik II. (1. Halbband)* (Zollikon, Evang. Verlag, 1945).
Rauschning, H., *Die Revolution des Nihilismus; Kulisse und Wirklichkeit im dritten Reich* (Zürich, Europa Verl., 1938).
Rosenberg, A., *Der Mythus des 20. Jahrhunderts; Eine Wertung der seelisch-geistigen Gestaltenkämpfe unserer Zeit* (München, Hoheneichen, 1930).
Rougemont, D. de, *Journal d'un intellectuel en chômage* (Paris, Albin-Michel, 1937).
—, *Journal d'Allemagne* (Paris, Gallimard, 1939).
Rousseau, J. J., *Du contrat social.* Présentation de H. Guillemin (Paris, Union générale d'éditions, 1963 (Le Monde en 10/18).
Russell, B., *History of Western Philosophy and its Connection with Political and*

Social Circumstances from the Earliest Times to the Present Day (London, Allen Unwin, 1946).

Ruyer, R., *L'Utopie et les Utopies* (Paris, P.U.F., 1950).

Saint Augustine (transl. J. Healey), *The City of God (De civitatie dei)* (London, Dent, 1945) (= *Everyman's Library*, no. 982).

Sarkisyanz, E., *Russland und der Messianismus des Ostens; Sendungsbewusstsein und politischer Chiliasmus des Ostens* (Tübingen, Mohr, 1955).

Schärer, H., *Die Gottesidee der Ngadju Dajak in Süd Borneo* (Leiden, Brill, 1946).

Schumpeter, J. A., *History of Economic Analysis* (New York, Oxford U.P., 1954).

Smith, A., *An Inquiry into the Wealth of Nations* (Dublin, Whitestone c.s., 1776).

Soekarno, *Sarinah; Kewadjiban wanita dalam perdjoeangan Republik Indonesia* (Djakarta, Pembangunan, 1951).

Spengler, O., *Der Untergang des Abendlandes. Umrisse einer Morphologie der Weltgeschichte* (München, Beck, 1927).

Thrupp, S. L., ed., *Millenial Dreams in Action, Essays in Comparative Study.* (= *Comparative Studies in Society and History: Supplement,* II) (The Hague, Mouton, 1962).

Troeltsch, E., *The Social Teaching of the Christian Churches,* 2 vols. (London, Allen Unwin, 1931) (2nd impr. 1949 used here). Original: *Die Sozial-lehren der christlichen Kirchen und Gruppen* (1911).

Wallace, A. F. C., "Revitalization Movements", *Amer. Anthrop.*, 58/2 264 ff. (1956).

Watt, W. M., *Muhammad at Mecca* (Oxford, Clarendon Press, 1953).

—, *Islam and the Integration of Society* (London, Routledge Kegan Paul, 1961).

Woodcock, G., *Anarchism; A History of Libertarian Ideas and Movements* (Harmondsworth, Penguin Books, 1963).

Wright, F. L., *The Living City* (New York, Mentor Books, 1958). (Ed. used here, 1963.)

Yang, C. K., *Religion in Chinese Society; A Study of Contemporary Social Functions of Religion and Some of Their Historical Factors* (Berkeley, Univ. of Calif. Press, 1961).

Bei Fragen zur Produktsicherheit wenden Sie sich bitte an:
If you have any questions regarding product safety,
please contact:

Walter de Gruyter GmbH
Genthiner Straße 13
10785 Berlin
productsafety@degruyterbrill.com